AF869105

The Musical Order of the World
Kepler, Hesse, Hindemith

" . . . the whole cosmos guided, ordered, and interpreted by the spirit of music."Josef Knecht in Hermann Hesse, The Glass Bead Game, p. 55 "the most profound order and magic of the laws as manifested in music, geometry, in the structure of plant and animal, the course of the stars, in essence, motion, relation, in all that has been created; which also includes the place, thought, deed, and destiny of human beings."Johannes Kepler in Paul Hindemith, Die Harmonie der Welt, act II "Die Sonne tönt nach alter Weisein Brudersphären Wettgesang" [The Sun, in ancient guise, competing With brother spheres in rival song]Johann Wolfgang von Goethe, Faust, "Prologue in Heaven," 1-2 "Die musikalischen Verhältnisse scheinen mir recht eigentlichdie Grundverhältnisse der Natur zu sein." [The musical ratios seem to me quite actually to be the fundamental ratios of nature.]Novalis, The Encyclopedia, IV/1, Language and Music, 1326

The Musical Order of the World
Kepler, Hesse, Hindemith

by Siglind Bruhn

INTERPLAY SERIES No. 4

PENDRAGON PRESS
HILLSDALE, NY

Other Titles in the Interplay Series

No. 1 *Masqued Mysteries Unmasked: Early Modern Music Theater and Its Pythagorean Subtext* by Kristin Rygg (2000) ISBN 1-57647-073-3
No. 2 *Musical Ekphrasis: Composers Responding to Poetry and Painting* by Siglind Bruhn (2001) ISBN 1-57647-036-9
No. 3 V*oicing the Ineffable: Musical Representation of Religious Experience* by Siglind Bruhn (2001) ISBN 1-57647-089-1

This book is dedicated to Robert Kessler of Pendragon Press
with gratitude for his unwavering support
throughout the past ten years.

Library of Congress Cataloging-in-Publication Data

Bruhn, Siglind.
The musical order of the world : Kepler, Hesse, Hindemith / by Siglind Bruhn.
p. cm. -- (Interplay series ; no. 4)
Includes bibliographical references and index.
ISBN-13: 978-1-57647-117-3
1. Hindemith, Paul, 1895-1963. Harmonie der Welt (Opera) 2. Hesse, Hermann, 1877-1962. Glasperlenspiel. 3. Kepler, Johannes, 1571-1630. Harmonices mundi. 4. Music and literature. I. Title. II. Interplay (Hillsdale, N.Y.) ; no. 4.
ML410.H685B75 2005
782.1--dc22

2005019274

Table of Contents

Part IV — Astronomy

Part V — Metaphysics

Part VI — Cosmic Harmony and Human Experience

Part VII — Art as a Hymn to Universal Harmony

Acknowledgments

As always, there have been many silent helpers looking over myshoulder during the genesis of this study. The staff at the Hindemith-Institut in Frankfurt kindly facilitated my exploration of the composer's notes and sketches—in music as well as, extensively, in words—on Kepler and his search for universal harmony. Bryan Skib and his colleagues at the University of Michigan's Harlan Hatcher Library have gone out of their way to identify and locate less common documents, and UoM's interlibrary loan office has been admirably efficient and fast.

Prof. Adrian Hsia of McGill University in Montreal, Canada, one of theworld's leading Hesse scholars whose own works show a particular interest in Hesse's interpretation of Chinese culture, generously read all segments pertaining to the novelist and commented graciously. Prof. Magnar Breivik of NTNU Trondheim, Norway, a devoted Hindemith scholar, was so kind as to peruse the entire manuscript and offer thorough comments, which helped me to clarify a number of hazy points and prevented some embarrassing errors from persisting into the printing stage. Prof. Gerhold Becker of Assumption University in Bangkok, Thailand, professionally a philosopher and ethicist and privately my best friend and husband of more than 20 years, helped me to substantiate my understanding of the Pythagorean background of the quest for harmony and contributed thoughts and probing questions onmany other aspects of the book, all along the way.

A sampling of my thoughts on Hindemith's Kepler opera has recentlyappeared in German ("Keplers Gedankenwelt in Hindemiths Musik,"*Hindemith-Jahrbuch / Annales Hindemith* 2004/XXXIII: 54-122). Anotheressay, also in German, on the role of music in Hesse's *Glass Bead Game*("Die Macht der Musik in Hermann Hesses *Glasperlenspiel*"), has been accepted by the editors of the journal *Musik & Ästhetik* for publication in the fall of 2005.

February 2005
Siglind Bruhn

Preface

In the disastrous years before and during the Second World War, when ever more aspects of life were overshadowed by Nazi atrocities and trust in a harmonious future was as difficult as it was crucial for spiritual survival, two German artists in exile wrote what would become their late masterpieces. The composer Paul Hindemith (1895-1963) collected material for an opera whose subject matter, the famous astronomer Johannes Kepler's mature life and theories, he had been considering since 1939. *Die Harmonie der Welt* [The Harmony of the World] was completed in 1956 and premiered at Munich's opera festival in August 1957.[1] The poet and novelist Hermann Hesse (1877-1962) conceived his "tentative description of the life of the *magister ludi* Joseph Knecht"—thus translates the original subtitle of *The Glass Bead Game*—in 1931, completed it in 1942 and first published it in Zurich in 1943. Both works address the topic of universal harmony in the fabric of creation and culture, as well as the urgent problem of how such harmony can heal the spiritual, mental, and emotional developments of individuals and of society at large. The two quests are mirrored into circumstances that are almost equidistant from the mid-20th-century period in which their stories are being told: Hindemith's opera centers on an outstanding intellectual hemmed in by the aristocratic whim and religious bickering rampant in the late 16th and early 17th centuries; Hesse's work focuses on this intellectual's counterpart projected into a fictional world of the early 23rd century, a world in which mental restrictions are subtle and decisions about life's course are made by that hierarchical society's steering leaders. True spiritual qualms, however, are painfully one's own in either scenario. In both cases, the quest for "harmony" and "truthful proportion" manifests at all levels of the stories told and of the works telling them.

The study is organized along the lines of the seven areas in which scholars of the Pythagorean tradition from Plato to Kepler and beyond found universal harmony paradigmatically realized: music, arithmetic, geometry, astronomy (which together formed the so-called *quadrivium* of the medieval

[1]Throughout this study, I will be using the key concept, "the harmony of the world," in three languages: the Latin *Harmonices mundi* for Kepler's treatise, the English *The Harmony of the World* for the translation of that treatise, and the German *Die Harmonie der Welt* for Hindemith's opera about Kepler.

liberal arts), complemented by metaphysics, psychology, and art (as an alternative to the classical *trivium*—grammar, rhetoric, dialectics—in the context of modern studies in the humanities). In each of the seven parts, the thought process begins with a reflection on the main insights developed over the millennia, and then proceeds to observations regarding corresponding phenomena in the two 20th-century works, Hindemith's opera and Hesse's literary text.

In the course of these observations, I will show that, shot through the seemingly simple weaving in seven aspects, the composer reflects on Johannes Kepler's scientific tenets, creative speculations, and poetic products not only by including them in his dramatic plot, and the novelist does not restrict his metaphor of an intellectual-spiritual harmonization exercise called the "glass bead game" to the content level of his narrative. To create a *dramatis persona* named Johannes Kepler after an historical person, and to fashion words that capture essential situations, thoughts, and traits of character, is a literary task. The way in which Hindemith sets such a drama as an opera, the means by which the music plumbs the depth of the subject matter in a process of interpretative transmedialization, is truly intriguing.[2] Similarly, Hesse bases his work on a layered literary structure, which consists of a fictional future chronicler's introduction, his biography of an allegedly famous person from his past, and this person's posthumous writings. These components are held together by a symbolic game and a transtemporal life. The musical structure of the whole and the themes and motifs addressed in the central Life as well as in its protagonist's own poetry and prose ultimately become functions of, or components in, this game.

In the case of *Die Harmonie der Welt*, the music is entrusted with an active role. As I will show in detail, Hindemith bases his compositional features—from the overall design through many intermediate levels of organization to a wealth of individual parameters—on the scientific and spiritual content of Kepler's writings. I will highlight in particular those aspects that have been unduly neglected in previous studies of the opera: the overture's

[2]Elsewhere, I have introduced the term "transmedialization" as a generic description for what in the field of comparative arts is called "ekphrasis": the re-presentation in *one* medium of a content previously represented in *another* medium. Thus Wallace Stevens's *The Man with the Blue Guitar* transmedializes Picasso's *The Old Guitarist* in a process in which poetry responds to painting, Schoenberg's *Pelleas und Melisande* takes its inspiration from Maeterlinck's symbolic drama, which he re-presents in instrumental music, and D'Annunzio's *Sopra un Adagio di J. Brahms* is a fine example of a work of music reflected in a poem. For a thorough discussion of the breadth of music's role in this fascinating arena see S. Bruhn, *Musical Ekphrasis: Composers Responding to Poetry and Painting* (Hillsdale, NY: Pendragon Press, 2000), especially pp. 3-104 and 561-586.

cosmological message; the opera's overall design with its hidden but spiritually evocative analogies and the scientist's original poem which this design mirrors; the relationship between some of the cosmic phenomena Kepler described and their analogs in the structural play of the five operatic acts; the composer's exploration of a piece of Keplerian juvenilia and its suggested impact on his family in the central act; Kepler's religious attitude, which Hindemith probes by means of numerous explicit and indirect musical quotations as well as through a play with symbolic numbers; and finally, the astronomer's cosmographically couched praise of the Creator, depicted in the opera's hymnic passages with a wealth of hermeneutically charged components.

Similarly in Hesse's *Glass Bead Game*, the structure, the thematic threads, and the narrative motifs within the work's component sections do much more than merely provide the scaffolding on which the story is erected. I concentrate above all on five aspects that have so far been underemphasized in the vast body of Hesse scholarship. Thus I demonstrate that the dimensions of the fiction are designed in symbolic consonance with one another, whereby all explore facets of a game that must be imagined as an ideational model hovering behind the various versions of the exercise that remains so tantalizingly elusive. What appears as an amalgam of four parts of different length, style, and apparent literary purpose is actually designed as a double-helix whose two strands reflect the focal issues in the dual title. Music does not only feature prominently in the narrator's stories but also informs the deep structure of one of the two strands. The only glass bead game about which readers learn enough to feel close to imagining its thematic content is based on principles that, although Chinese and thus ostensibly far from the Pythagorean tradition, match almost all aspects in the Western quest for universal harmony. And while the glass bead game is a symbol of Hesse's utopian World of the Mind, the Glass Bead Game Master portrayed in the central Life is, in his most significant aims and his life's trajectory unto death, himself in turn a symbol of the glass bead game.

My introduction fulfills three purposes. It begins by reminding readers of the tradition of Western thought about "universal harmony," tracing its main stages in lines that converge on the person of Johannes Kepler. This is followed by sections focusing on Hesse and Hindemith respectively, which, among other data, provide the fundamental details about the genesis and compositional structure of the two works investigated.

One facet, although constituting significant background information, is not included in the introduction. It will become obvious in the course of the study that while Hindemith's work is rooted exclusively in the Western tradition of thought about universal harmony (with its scientific, aesthetic, ethical, and metaphysical aspects), Hesse's work draws additionally on ancient Chinese thought. The scope of this study would not have allowed to attempt a developmental sketch of that exceedingly rich and complex tradition. I have therefore limited myself to three aspects. One, which is simply indispensable for an understanding of Hesse's aims, is the wisdom and oracle teaching of the *Book of Changes*, which I discuss in part 5. Another aspect informs the particular glass bead game conceived by the *magister ludi* at the height of his career; this is summarized in part 3. Finally there is the traditional Chinese attitude to music.

To this day, practising Daoists credit music with magical influences on humans and nature, while Confucians believe that music has moral effects, that it is linked to and capable of mirroring socio-political situations. Even older teachings combine the two stances, considering music the original fountain of order, morality, beauty, and health. Since this aspect is so germane to the title and aim of my study, I have decided not only to discuss (in part 1) the ideas Hesse quotes from a book that purports to compile Chinese thought of the times before Confucius and Laozi, *The Annals of Lü Buwei*, but have actually copied from the English translation of this work the most relevant excerpts on music. These appear in the appendix.

Kepler: Music and the Quest for the Harmony of the World

Harmonices mundi: A Treatise about Universal Consonance

In 1618, Johannes Kepler, imperial mathematician in Prague, completed a five-volume treatise on what he described as "the harmony of the world" (*Harmonices mundi libri V*). The work excited astronomers owing to the final law of planetary motion, which complemented Kepler's earlier findings to the now famous threesome: (1) planets move in elliptical orbits, with the sun at one focus; (2) the radius vector from the sun to the planet sweeps over equal areas in equal times (in other words, the closer a planet comes to the sun, the more rapidly it moves); and (3) the square of a planet's orbital periods is proportional to the cube of its mean distance from the sun.

These laws, which Kepler developed on the basis of data collected by the Danish astronomer Tycho Brahe, may constitute a giant step for humanity's grasp of what they see occurring in the sky, yet they are but a minute portion of the argument. As Kepler emphasized many times, the laws are a mere tool and as such subservient to his nobler goal: to demonstrate the identity of order in nature, in the psycho-physical disposition of humans, and in music; to prove that the age-old belief in a universe resonating in accordance with musical consonances can be corroborated with the help of 17th-century physics; and to show that there are reasons to trust that the same harmonious proportions pertain to innumerable other aspects of the universe.

Kepler laid out his study as follows: In Book I, "Geometry," he shows how the regular figures that manifest harmonic proportions are developed. In Book II, "Architecture," he explains how the regular plane-geometrical figures can be joined to form solids. In Book III, "Harmony," he demonstrates the relationship between the laws of music theory and proportional harmony in regular figures. In Book IV, "Metaphysics, psychology, and astrology," he discusses the spiritual nature of harmonies and their manifestations in this world, above all the harmony created by the rays of the heavenly bodies shining upon the earth, and its impact on nature, the World-Soul, and the human soul. In Book V, "Astronomy and metaphysics," he draws on the laws of harmonic proportion to explain the range of each planet's orbital speeds along with the ratios of the fastest and slowest speed values of adjacent cosmic bodies. In an appendix, Kepler compares his study with Ptolemy's *Third Book on Harmony* as well as with Robert Fludd's harmonic deliberations on macro- and microcosm.

As this overview shows, only Book V of Kepler's extensive treatise focuses on planetary motions. The reason why the term "the harmony of the world" is widely misunderstood as referring primarily if not exclusively to events in the skies—with the popular notion that it describes a kind of astral concert—is at once etymological and historical. The historical reason goes back to Pythagoreanism, which is reputed of having centered on a knowledge of celestial music. (This school was fashioned after the ancient mystery religions whose "esoteric" wisdom was never shared outside a small circle of initiates. Thus very little of their belief and practice is certain.) In a passage of his *Republic* whose detailed specifications scholars take for a veiled esoteric message in the Pythagorean tradition, Plato describes singing sirens sitting on the planetary spheres.[3] At about the same time, Aristotle explains that the planets can only generate their tones, which humans do not hear, because of the wise way in which the proportions of their orbits' relative distance and speed have been arranged.[4] The tradition of a quest for "the music of the spheres," which extends through Ptolemy's speculations and beyond, perpetuates the limiting notion that in the phrase "the harmony of the world," the "world" means the cosmos.

The other reason for the misunderstanding lies in Kepler's key term, the word *harmony*. (*Proportion* and *consonance* are sometimes used as its synonyms; they share the comprehensive denotation, which is not restricted to any specific field of application.) The word ἁρμονία is of Greek origin; its etymology combines the initial syllable *ar* or *har*—which, in the Indo-Germanic languages, denotes the unification of contrasts or strong differences to a reconciled whole—with the suffix *-monía* signifying action, state, or condition. Harmony, then, is the action or condition by which contrasts or differences are unified. At the earliest stage of its usage, the term seems not yet to have been connected with numerical laws.

[3]See the story Socrates tells in Book X of his *Republic* about Er the Pamphylian. A hero slain in battle, Er had what our age would call a near-death experience, in which he was privileged to see the next world before returning to life. Er's vision includes a model of the universe as a set of concentric whorls (the planets) hung on the spindle of Necessity. The rims of these whorls, of different sizes and colors, revolve at different speeds—all the inner ones in opposite direction to the outermost one, the firmament.

[4]See *On the Heavens*, Book II, Part 9.

The Development of the Pythagorean Concept of Harmony

Harmony developed only gradually from a mythological to an allegorical figure and from there to an abstract concept. In ancient Greece, Harmony was a goddess in the Olympian pantheon. In the mythology of the peoples from Boeotia and Illyria, she was known as the daughter of Ares, the god of war, and of Aphrodite, the goddess of beauty and love. The Romans adopted this genealogy and made *Harmonia* the daughter of Mars and Venus. A similar ancestry is known from Babylonian, Persian, and Germanic mythologies, where a corresponding figure was thought to bridge the contrasts of light and darkness, good and evil, or heaven and hell. Moreover, the unification of opposites in a dualistically conceptualized reality was applied to the macrocosm as well as to the microcosm, to the implications of stellar positions and the lives of humans. A different strand of Greek mythology, found in Attic tales, focuses on Harmony's daughters, the muses, through whom she became the protector and keeper of the humanities and the liberal arts.

In the more philosophical circles, Harmony mutated to an allegory: the goddess whom the common people perceived as a personal power bringing order into the world, was here understood as a symbol and, eventually, as a concept. While this concept did not gain currency until the era of Greek Classicism, it may be as old as the goddess; it is generally traced back to the Pre-Socratic philosophers, especially to Pythagoras and his followers.

This extraordinary philosopher and religious teacher's intellectual and spiritual background was essential in shaping the idea of "universal harmony." Born on the Greek island of Samos in the early 6th century B.C.E., Pythagoras was a contemporary of all the seminal thinkers that shaped the understanding of religion and philosophy throughout the ancient world: his life span (he was born ca. 570 B.C.E. and died in 497 or 496) falls between those of Zarathustra (ca. 630-553), Confucius (ca. 551-479), and the Buddha (ca. 560-480). He spent his formative years in Miletus, Egypt, and Babylonia, studying the sciences for which each school was famous, above all arithmetic, music, geometry, and astronomy. In all three centers of learning, he was also initiated into the religious rites and teachings.

Having briefly returned to Samos but finding it subdued by the tyrant Polycrates, he emigrated (probably around 529) to Crotona in southern Italy, a port 175 miles southeast of Naples. Here he founded a school and brotherhood that lived according to strict precepts. These concerned the members' religion, livelihood, philosophical stance, political engagement, and scientific inquiry. Pythagoras regarded these various fields as strongly interconnected. He was convinced that an understanding of the laws of arithmetic and geometry, the exploration of musical consonance and the motions of heavenly bodies, and finally the recognition of corresponding harmonic laws in all created nature served to purge the soul, benefit its concordance with the cosmos, further its liberation from the circle of rebirths, and prepare it for its journey into heaven.

Although the Pythagoreans were attacked, for the first time still in their founder's lifetime and repeatedly thereafter, the adherents to these teachings lived on in various communities and preserved their thinking throughout the 5th and 4th centuries B.C.E. In their continued study of the exact sciences as well as in their conduct, they were guided by a belief in the importance of numbers as the essential tool for an interpretation of the world. As recent scholarship confirms, many of Plato's ideas draw on Pythagorean concepts of harmony. This is particularly true of thoughts exposed in *The Republic* and *Timaeus*, a dialogue named after a 4th-century B.C.E. historian and Pythagorean chronicler.

Neo-Pythagoreans include Greeks and Italians living as late as in the first centuries C.E. Influential among them is the mathematician Nicomachos of Gerasa, whom Boethius will later quote extensively. Like the founder generation centuries earlier, the Neo-Pythagoreans believed that the planets are deities;[5] that their motions and constellations therefore have an impact on everything in the cosmos, including the human soul; that the soul is immortal; that souls incarnate in various (not always human) bodies; that all sentient beings should therefore be considered as related with one another and treated with respect; and that everything that has ever come about will recur in continuous cycles.

In Kepler, the Pythagorean world view as designed in innumerable manifestations of universal consonance found a unique interpreter. Kepler revered Plato, whom he quotes and discusses extensively, primarily for his insights into the metaphysical components of harmonic observations. Plato spelled out what Pythagoras and his contemporaries almost certainly believed as well: that harmony is based on numbers, usually in the form of ratios that stand for intervals; that these intervals can be experienced directly by the ear and, through it, by the soul; and that mathematics should therefore be understood as an ontological mediator between ideas and phenomena.

Plato believed in a development from the realm of the spiritual to that of physical being, along the sequence: number / line / plane figure / body. This

[5]The concept of planets as deities is reflected in our weekdays, named after the sun (Sunday, Sonntag) or the lord of gods, *dominus* (It. *domenica)*, the moon (Monday, Montag) or *Luna* (Fr. *lundi*, It. *lunedi*), Mars (Fr. *mardi*; for "Tuesday" see the corresponding Norse war god, Tiu), Mercury (Fr. *mercredi*; "Wednesday" after the Norse messenger god, Wodan), Jupiter (Fr. *jeudi*, It. *giovedi*; "Thursday" after the Norse thunder god, Thor), Venus (Fr. *vendredi*, It. *venerdi*; "Friday" after the Norse goddess Freya), and Saturn (*Saturday*).

graded ascent passes from one dimension to another in so subtle a way that it counteracts the imagined dualism of mind/spirit and body. Not surprisingly therefore, Plato's World-Soul is fashioned along mathematical lines that veil a musical reality (see part V below).

Three thinkers constitute the main links in the chain that connects Kepler to the Pythagorean school: Ptolemy, Boethius, and Nicolas of Cusa. Claudius Ptolemaeus (ca. 83-161 C.E.) was a Greek mathematician, geographer, astronomer, and music theorist who spent his working life in the Egyptian city of Alexandria, where he probably had access to the famous observatory. His major work, on mathematics and astronomy, the *Almagest* (Greek title *Mathematike syntaxis*), compiles his era's knowledge of the universe. Ptolemy also wrote what is considered the most learned, systematic, and comprehensive exposition of music theory in antiquity, *Harmonika*. In it he discusses the theory of harmonics (overtones), the principles of acoustics, interval theory (with a critique of the theory taught in the Pythagorean and Aristoxenian schools), and the theory of the genera, the various divisions of the tetrachord. He also describes the monochord and the helicon, two instruments used not for musical performance but for an appreciation of harmonious proportions, and compares musical tones with parts of the human body and with planets.[6]

Ancius Manlius Torquatus Severinus Boethius (480-525) transmitted the ancient knowledge to the Middle Ages. His writings on music and arithmetic, especially his five-volume *Fundamentals of Music*, provide a good overview of ancient music theory in its practical as well as speculative dimensions. In book I:2 he explains music's threefold manifestation as *musica mundana*, *musica humana* and *musica instrumentalis*—where *musica mundana* refers to music as an all-pervading force in the cosmos, *musica humana* to music as a principle unifying the human body and soul, and *musica instrumentalis* to music as sung and played by humans: concepts that reflect Pythagorean ideas. In book II:20, building on the arithmetic of Nicomachus, Boethius revives the use of the *lambda* figure (Λ) to illustrate the analogy of the series $^1/_2$, $^1/_3$, $^1/_4$, $^1/_5$, $^1/_6$ and $^2/_1$, $^3/_1$, $^4/_1$, $^5/_1$, $^6/_1$. This figure will be revived in modern thinking about universal harmony.

In the late Middle Ages, Nicholas of Cusa (1401-1464) explained the relationship between the finite and the infinite with the help of mathematical correspondences. Defining God as the *coincidentia oppositorum*, as one who encompasses everything, including all opposites, he stressed a thought that harkens back to Pythagoras and Plato: that the world's beauty is owed to

[6]For an accessible and lucid discussion of Ptolemy's importance for early concepts in music theory, see Lukas Richter's article in *The New Grove Dictionary of Music and Musicians*.

God's having created it on the basis of the numerical proportions obtaining in music, i.e., in correspondence with the ratios that define consonant intervals. Nicholas, too, understands harmony as the inherent condition of all that is divinely created. The integrity of all aspects of this world grants the unified order of what, he stresses, is not haphazardly called the *uni*-verse. The study of the musical condition of all things and species forms part of the essential knowledge of God; grasping harmonic principles is the first step toward insight into cosmic consonance.

These were Johannes Kepler's main ancestors. He perceived the inquiry into proportional harmony like them: as a method of thinking in analogies. The Pythagoreans taught the analogy of mathematical (and thereby, harmonic) proportions: e.g., 2 : 4 = 3 : 6. Plato employed the idea of analogy in a more general sense when, in his *Timaeus* and *The Republic*, he related the world of ideas analogically to the world of matter. Medieval Neo-Platonism held that music is an analogy to all ordered being. In this tradition, Kepler undertook morphological comparisons of specimens or aspects taken from different realms. Proving their identical structures, he showed that all aspects of nature are governed by corresponding laws. Human beings are susceptible to universal harmony on two levels. Primarily and generally, they perceive it on a subconscious level; secondarily and selectively, they can become consciously aware of it by studying its manifestations and expressing their observations in numerical ratios. When these are translated into musical intervals, the human ear's innate affinity with perfect proportions allows us to sensually confirm the result as harmonious. From this understanding derived the ancient discipline of *musica speculativa*, in which music was understood to serve as a mirror (Latin *speculum*) for the remainder of the world. Consequently, the first postulate of speculative music is that tones, like numbers, are ontologically prior to material existence.

In this tradition of thinking, in which Renaissance Europe joined many non-European cultures, music was revered both as an archetype of material order and as an analogous expression of spiritual harmony. The French jurist and natural philosopher Jean Bodin (1530-1596) exemplifies this attitude by means of an anecdote. In *The Colloquium Heptaplomeres*, he describes a dispute between representatives of seven faiths: pagan, Jewish, Catholic, Muslim, Lutheran, Calvinist, and Naturalist. The men are gathered to identify a truth on which they can be united. "At the close of each day's conversation, the participants are entertained by boys performing sacred music with voices and instruments, in a ceremony that restores harmony between the seven friends, no matter what their temporary disagreements may have been. The entire Colloquium is like a work of musical polyphony in which the different voices combine, both in concords and discords, to make a harmonious whole that is true to Nature itself."[7] 300 years later, the German lawyer and philologist Albert von Thimus (1806-1878) pursued a similar goal in his extensive treatise *Die harmonikale Symbolik des Altertums*. Developing a claim Georg F. Creuzer had made in a study of the symbolism and mythology of ancient peoples—namely, that all mythologies have a common source—Thimus suggests that the shared inspiration consists most probably in the independent discovery of the harmonic series. He demonstrates this view with abundant evidence from Chinese, Indian, and Egyptian writings as well as from works of Greek philosophy and the Church Fathers. In his wake, the Swiss music theorist Hans Kayser (1891-1964) designed a modern theory of harmony which, on the basis of Pythagoras's insights, aimed at complementing the visual with an aural appreciation of universal consonance (*aesthesis* with *acroasis*).

In the modern era, analogous modes of thinking have largely been ousted by scientific modes. These stress the "efficient cause" over a number of other potential causes (*causa formalis, causa materialis, causa finalis*), which in ancient and medieval cultures were considered at least equally important for understanding nature in both the narrow and the wider sense. Recently, information theorists think in analogies, as do biologists (if rudimentarily) when they compare the DNA code with the identical "code" used by the Creator of the universe.

Proportional harmony—the harmonious relationship between the parts, with and within the whole—can be sensually verified above all in sounds. But it occurs equally in the abstract realm (arithmetic, geometry), in nature (plants, human and animal bodies, crystals, atoms, electrons, molecules, planets, etc.), in art (musical compositions, dance, literary forms, paintings, sculpture, architecture, etc.), and in speculations about social, philosophical, and metaphysical truths (the human soul, the World-Soul, ideal communities). Some of these areas will be briefly explored as foils against which Hesse's literary work and Hindemith's opera can be highlighted.

In a work of art, harmony can manifest itself in content and structure. It can determine the nature of the story that is being told, the subject matters discussed among the protagonists, the way in which the author in turn shapes the telling and dramatizing, and various other aspects. Exploring Hesse's and Hindemith's works from each of the aspects defined by the age-old tradition of research into universal harmony leads to unexpected new vistas and exciting discoveries.

[7]Quoted from Joscelyn Godwin, *The Harmony of the Spheres: A Sourcebook of the Pythagorean Tradition in Music* (Rochester, VT: Inner Traditions International, 1993), p. 215.

The Life and Time of the Future Operatic Protagonist

Johannes Kepler (1571-1630) lived in a time of mental and military turmoil, between Renaissance and Reformation on the one hand, Counter-Reformation and the Thirty Years War (1618-1648) on the other. In the decades preceding his birth, the rediscovery of several of Plato's works had given new impulse to philosophy and spawned new approaches to the thinkers of ancient Greece. Martin Luther's translation of the Bible into the German vernacular had expanded the attitude of making the past speak to the present by granting the educated lay person direct access to the sacred text of the Judeo-Christian heritage. Finally, the invention of printing with move-able type made scholarly and biblical texts accessible to readers outside monasteries and learned communities.

At the time of Kepler's youth, however, zealous Protestants began to have qualms about the pre-Christian, "pagan" components of Greek thought, and eventually reduced ancient philosophy to only those elements that are compatible with the Christian message. In Kepler's later years, scientists developed a bias toward quantifiable data and began to favor a rationalism and empiricism that would segregate physics from metaphysics, natural science from philosophy and theology. Engineering, an applied discipline that did not form part of the liberal arts, had only recently gained importance in the context of new developments in weaponry. Now Galileo Galilei's ballistic experiments as well as the invention of the watch, the telescope, and the microscope seemed to make it imperative that the gap between academic scholarship and applied technology be bridged. Numbers, which had always been treated as both rational entities and mystical symbols, were stripped of their spiritual dimension to become the essence of all scientific inquiry. Even more disturbingly, Copernicus's discovery that the planets moved not around the earth but around the sun forced humankind to rethink the importance of their planet—and thus of their own position—within the cosmos. As the centrality of the human world in the universe seemed no longer a given, the question arose whether God had indeed created the world for the sake of the human race.

Religious strife also took a new form. Luther's eagerness to reform the Church had divorced Protestants from Catholics. Now competing reformers, most notably Zwingli and Calvin, disputed individual points in Luther's teaching, with the result that the various factions of the Reformation move-ment soon fought more fiercely against one another than against the Roman Church. The Vatican responded with the confirmation of the Jesuit Order, whose primary goal it was to restore the "right" faith and thus, to fight all Protestants. The resulting Thirty Years War led to much bloodshed and lasting enmity. As a side effect, it also raised the question who should have the final say on issues of faith: the State or the Church. The divergence of opinions in turn initiated the creation of absolutist nation states.

Kepler was born into a Lutheran family in Weil der Stadt, a little town in what is now the south-western corner of Germany. His father, a dissatisfied and adventurous man, hired himself out as a mercenary to Catholics and Protestants alike and eventually disappeared, leaving his wife and children in dire straits. Kepler's mother was the dominating presence during the first twelve years of his life, until he left home for his designated education.

Kepler's schooling, determined by his great intellectual ability in combination with the poverty of his home, allowed only one route. After an initial year at a local school that taught reading and writing in German and three years at a local "Latin school," he passed the centralized examination on the basis of which the Lutheran Church selected future clergy. The successful candidates were given grants and assigned to boarding schools in former monasteries that had been refashioned in terms of doctrine but not with regard to the severity of the students' everyday life. After four years, aged 17, Kepler received the Bachelor's degree; having completed the required additional formative year, he was admitted to the theological seminary at Tübingen University. Here he studied for two years for a Master's degree in the liberal arts before beginning his final three-year course in Lutheran theology.

A few months before the conclusion of these studies, Kepler's teachers nominated him for a position as professor of mathematics at a sibling institution, the Lutheran seminary in Graz. The choice of this particular student may have had two reasons. Kepler had shown extraordinary abilities in all fields of abstract and applied mathematics and was thus intellectually particularly well suited. At the same time, his future as a theologian was seriously jeopardized by the fact that he had developed doubts about the Lutheran interpretation of the Eucharist, which stressed the *real* presence of Christ's body and blood, while Catholics and Calvinists believed this presence to occur *in spirit.* Although Kepler was aware that dissent on even a single point of doctrine would not be tolerated, he felt unable to dissimulate and had confessed his doubts to one of his teachers. Despite near-completed studies in theology, he would thus have been untenable as a clergyman.

Kepler spent six years in Graz. In 1597, he published his first treatise, the *Cosmographic Mystery*, and married. By 1600, the region around Graz had succumbed to the forces of the Counter-Reformation and all Protestants were expelled. Kepler moved to Prague, where the reputed astronomer Tycho Brahe (1546-1601) found his mathematical skills useful and hired him as his assistant. Upon Brahe's death, Kepler succeeded him as imperial mathematician at the court of Emperor Rudolf II. He continued to publish, now especially in astronomy and astrology but also combining these fields with his background in theology (as when he offered astronomic arguments for determining the true year of Christ's birth).

In 1612, Kepler's life was uprooted once again. A year earlier, his wife had died, following the third of five children from this marriage. Early in 1612, Emperor Rudolf had also passed away. Although Emperor Matthias confirmed the post of imperial mathematician, Kepler's salary was rarely paid and he was forced to seek other means of income. Thus he accepted a position as "mathematician and cartographer" offered by the guilds of Linz, in Upper Austria. Hardly had he begun work there than the local Lutheran minister revealed that he had learned of Kepler's dissenting view on the Eucharist and required that Kepler sign a special confession confirming his conversion to the doctrine. When Kepler refused, on the basis that no lay person should be required to sign such a statement, the minister publicly announced that Kepler was henceforth excluded from communion and prevented from attendance at his church. This verdict was to pursue Kepler for the remainder of his life.

In 1613 Kepler remarried. With this second wife, Susanna, he would have seven children, four of whom were still alive at the time of his death. In subsequent years, he completed numerous essays on topics in arithmetic, geometry, astronomy, chronology, and theology as well as his magnum opus, *Harmonices mundi libri V* (1619).

In 1625, Protestants were expelled also from Upper Austria. Although Kepler was originally exempt from the expulsion, he and his family had to flee when battles and burnings reached the city of Linz. After three worrisome years of unemployment, Kepler had no choice but to accept a position offered him by the leading general in the army fighting for the Catholic Emperor, Count Wallenstein (1583–1634). This position, once again calling for a "mathematician," required him to calculate anything from army traveling time and geographic obstacles to supplies, as well as making astrological prognoses for the heroic general. In the midst of all these demeaning chores, Kepler succeeded in continuing his scientific research. He died in 1630, shortly after learning that the Emperor had discharged Wallenstein and he had to fear that his life and that of his family were once again unsupported. The general was murdered four years later by his own officers, in the very month beyond which Kepler had refused to expand his horoscope.

Kepler's advancement of physical science was a secondary effect of his primary concern with universal harmony. The breadth of his contribution to knowledge equals that of Isaac Newton, who considered his research into alchemy, biblical chronology, and prophecy fully as important as his work in physics, mathematics, and astronomy.

Hesse: In Search of Consonance in the World's Cultures

Castalia and Its Exemplary Student

The lifetime of Joseph Knecht, the fictional protagonist of Hesse's *Glass Bead Game*, bridged the 22nd and 23rd centuries—or so the reader learns from a chronicler allegedly writing around 2400.[8] In the course of the two centuries preceding Knecht's birth, Europe is reported to have painstakingly overcome the period of intellectual prostitution and spiritual degeneration that climaxed in the 20th century. By way of a reformation that focused not on the purity of religious faith but on the purity of the human mind, it has eventually restored dignity to scholarly endeavors while raising them to new heights.

Knecht's life is supported and defined by a structure that owes its determining features to the quagmire out of which it arose. That sad era is referred to as the "Age of the Feuilleton," a term allegedly coined by the leading 24th-century historian of the period. *Feuilleton*, a French noun used as a loan word in other European languages, is the heading of that section in newspapers that is made up of entertaining essays on cultural events.[9] If one can believe the analyses offered by the baffled historians of later centuries, the Age of the Feuilleton must have been marked by technological and ideological hubris, intellectual superficiality, and particularly by a leisure culture in which literary products were converted to a marketable commodity adjusted to the lowest common denominator of the half-educated bourgeoisie.

[8]This dating derives not from the work's text but from the remarks with which Hesse prefaced the preview publication of the chronicler's introduction in Suhrkamp's *Neue Rundschau.*

[9]How much Hesse considered this era—his own life time—as an aberration that must soon be overcome may be gleaned from the words with which, in the early 1920s, he opened his "Kurzgefasster Lebenslauf" [Concise Curriculum Vitae], an imaginatively commented autobiography complemented with a conjectural account of the second half of his life: "I was born around the end of the modern era, not long before the incipient renewal of the Middle Ages." Translated from the reprint in Volker Michels, ed., *Hermann Hesse, Sämtliche Werke* 12 (Frankfurt: Suhrkamp, 2003), pp. 46-63 [46].

According to the chronicler, there followed from the mid-20th century onward a period of transition shaped by "a process of self-examination, reflection, and conscious resistance to decline." A new seed emerged and subsequently evolved in three stages. Musicologists with their historical, theoretical, and aesthetic concerns came to the fore, offering a haven for scholars eager to preserve a core of good tradition and intellectual rigor. The sober attitude was spiritually complemented by the wisdom (and, later, the meditative practices) of the Journeyers to the East, many of whom were exemplary performers of early music.[10] There arose a new conception of music, characterized by awe and a resigned acceptance of the inevitable aging of cultures. In the course of many decades, this asceticism led to a reformation and the development of a "World of the Mind" within the State. A secular order was founded, whose members renounced assets, family, and self-determination.[11] Hierarchically structured, it built on a system of elite schools, institutes of advanced study, archives, etc. Its name, "Castalia," points to the Parnassian spring in which Greek pilgrims purified themselves before entering Apollo's temple at Delphi. With a bow to the institution to which Goethe entrusted his hero's son in *Wilhelm Meister's Journeyman Years*, Castalia is also referred to as "the Pedagogical Province."

Joseph Knecht, an orphan of unknown background, first distinguishes himself in the "Latin school"of a small Swabian town—one is reminded of Johannes Kepler, whose initials he shares—as a student particularly gifted in music and the Classical languages. Having been recommended to and subsequently examined by Castalia's Music Master, he is elected for education in the elite schools.[12] After four years, now almost 18 years old and ready to progress from the junior to the senior schools, the Board of Education assigns him to Waldzell. This is the school that breeds the glass bead game players, the elite of the elite. Their annual festive performance, attended by hundreds of visitors, is considered the outstanding aesthetic and

[10]This mysterious group, the dedicatees of the book, derive their name from a novella Hesse wrote shortly before embarking on *The Glass Bead Game*. The term embraces anybody who, committed to a life of the mind, is on a path toward self-understanding and purification. In addition to intellectual discipline, the brothers of this "league" stress spiritual purification.

[11]While its members—men only in this world!—remain single in order to devote themselves fully to their spiritual tasks, they are not, like their counterparts in Christian orders, held to chastity. We learn that the daughters of the townspeople tended to marry late and prided themselves of their premarital relationships with these friars of the mind.

[12]Most of these schools are said to be housed in former monasteries. Their models are precisely those schools for prospective Lutheran clergy in which Kepler received his education and from one of which the 14-year-old Hesse defected within months of being admitted.

spiritual event of every year. Having initially focused on music, Knecht later submits to the school's broad-based curriculum of arts, humanities, sciences, and meditation exercises, and by the time of his graduation at age 24 has turned into a promising glass bead game player.

During his Waldzell years, not all is smooth though. Debates with a guest student about the strengths and weaknesses of Castalia versus "the world outside" raise first doubts in Knecht whether the Order to which he aspires to belong is really all it should be. Not much later, a new Castalian friend, Fritz Tegularius (whom Hesse modeled after Friedrich Nietzsche), involuntarily exacerbates Knecht's misgivings. He is a brilliant scholar but a physically frail and emotionally unstable man. Knecht recognizes in him an example of the highly-strung intellectual who will never serve either the Order or the world at large in any capacity and thus embodies the extreme to which pure esthetics can lead. Considerably shaken by his unexpected qualms, Knecht vents his feelings in a series of poems, although such activity defies the Castalian ban on all artistic creation.

After graduating from Waldzell, Knecht is granted a 10-year period of free study. In order to check both their students' studiousness and maturation process, the Board of Education requires them to write several fictional autobiographies imagining themselves in past epochs. Beyond these tasks, Knecht is free to apply himself to whatever interests him. He chooses a strange task: to retrace the full historical, etymological, and symbolic background of a specific glass bead game that he once co-designed, a game whose two themes were a fugal theme and a saying by Confucius.

The latter theme gives Knecht occasion to further his studies in Chinese language and culture, which have long fascinated him. To this end, he spends several months with an atypical Castalian known only as "Elder Brother." This recluse, an accomplished Sinologist who has blended knowledge and life, object and subject to the point at which he himself has become indistinguishable from a Daoist hermit, serves as both a mentor in Chinese wisdom and a warning example about escapism, which Knecht recognizes clearly as not his lot.

At age 34, Knecht is inducted into the Castalian Order. As a bridge between his self-directed time of study and his ultimate integration into the hierarchy, he is dispatched on a diplomatic mission. For several years he is to live as a guest in a Benedictine monastery, where his official assignment is to give courses in the glass bead game to interested monks. Beyond this cover activity, he is to gain the abbot's trust and thereby further Castalia's hopes for mutual recognition and official intercourse between the two institutions. Knecht succeeds well: he wins the friendship and trust not only of the Benedictine abbot, largely on the basis of their shared interest in

Chinese wisdom and the *I Ching*, but also of the outstanding historian Father Jacobus (modeled after the Swiss historian Jacob Burckhardt [1818-1897], whom Hesse admired). The latter becomes his mentor in the study of historical thought, instilling in him a sense of the interdependence and transience of all human endeavors—even such endeavors of the mind as the Castalian project—while in turn allowing himself to be slowly won over to respect the young and churchless order.

After approximately four years at the monastery, Knecht is called back to Waldzell. Shortly after his arrival, the current *magister ludi* dies and Knecht, although not yet 40 years old, is elected to become his successor. During the subsequent eight years he is highly successful in all the administrative, diplomatic, and educational tasks connected with this position, and the ceremonial glass bead games he composes and celebrates are admired as master pieces in the history of the Castalian *ludus solemnis*.

But Knecht's doubts about Castalia's splendid isolation from the world-at-large persist and increase. He acknowledges the dangers inherent in intellectual sophistication pursued for its own sake—a one-sidedness that leads to caste arrogance and eventually renders its achievements irrelevant to real-life needs. Knecht reestablishes contact with the former guest student Plinio Designori, and the erstwhile debating opponents, who had felt estranged in intervening years when one became ever more refined and serene and the other ever more worldly and bitter, now meet once again on a formally equal footing: a high government official and successful political writer encounters a *magister* of the Order and brilliant glass bead game player. Their revelatory conversation convinces them of the distance of their respective worlds and of their shared desire for harmonization. As a result, Designori reconnects with his Castalian heritage and finds relief for some of his sorrows, while Knecht acts on his wish to pass beyond the borders of the World of the Mind and serve outside, in the hope thereby to achieve a larger harmony for himself and for the Castalian idea.

In a circular to the Castalian directorate Knecht asks to be assigned a position as a simple schoolmaster, knowing full well that such a wish can never be granted. Upon receiving a negative reply and remonstrance, he takes the step that is unheard-of in higher officials: he resigns and leaves all that has so far been his life. Designori has offered him to tutor his gifted but difficult son. But hardly has Knecht made progress in winning the skeptical youngster's trust, than he dies in a swimming accident—leaving his pupil in awe of the joint legacy inherited from father and teacher, natural world and Castalia, which it will be his life's task to harmonize.

The Glass Bead Game: Genesis and Structure

Hesse's first sketch for his late masterpiece dates from 1931.[13] It shows that he originally planned a work based on the fivefold rebirth of a man tentatively named X: (1) as a rainmaker in a matriarchal culture, (2) as a hero who founds a world empire; (3) as a Christian knight or monk; (4) as a man of the present, who tells the saga of X; and (5) as a "bead player" in a future age. Of the three lives included among Knecht's posthumous writings, "The Rainmaker" and "The Father Confessor" are easily identifiable as corresponding to numbers 1 and 3, and one recognizes the "bead player" in the protagonist of the Castalian biography. The second incarnation is too vague in the sketch to be matched with any particular story; its reference to the empire may point negatively to the Maya dream in the "Indian Life," where a splendid kingdom is lost. The contemporary life alluded to as no. 4 seems to have ended up split in two. One half can be ascertained in the chronicler's account of the feuilletonistic era—clearly intended to describe Hesse's 1930s, but still eerily valid for our early 21st century. The other half turned into a slightly postdated variant, the (unfinished and separately published) life of an 18th-century Swabian theologian and organist.

Having drawn up this plan, Hesse soon composed several self-contained segments: the four versions of the "introduction to the glass bead game" (between 4/1932 and 6/1934), twelve of Joseph Knecht's poems (12/1932-11/1937), the fictional lives (1/1934-4/1937), the chapters of the main Life (1/1938-4/1941), and the poem "Stages" as a final thought in 5/1941. All poems were published in journals shortly after their completion, as were the lives, the introduction, and seven of the twelve *magister ludi* chapters—most of them in Suhrkamp's *Neue Rundschau*, some in the Swiss literary journal *Corona*. The entire work, which Hesse's faithful German publishing house Fischer attempted to bring out in 1942, was rejected by the Nazis' cultural office but appeared in 11/1943, published in Zurich by Fretz & Wasmuth.

The work is generally described as consisting of three segments. This design seems to be corroborated by the table of contents, whose main headings are "The Glass Bead Game: A General Introduction to Its History for the Layman," "The Life of the Magister Ludi Joseph Knecht," and "Joseph Knecht's Posthumous Writings." However, the segments do not relate to one another in a linear order, as the verbal sequence might seem to suggest.

First, there is the motto: supposedly an excerpt from a "Treatise on the spiritual value of glass beads" written by one Albertus Secundus in scholastic Latin and given with a German translation by none other than Joseph Knecht

[13]The sketch is reprinted in Volker Michels, ed., *Materialien zu Hermann Hesse 'Das Glasperlenspiel'* (Frankfurt: Suhrkamp, 1973), vol. I, pp. 314-315.

himself. The 68-word prose text relates to the main Life in an indirect way, insofar as the famous *magister ludi* apparently found it worth his attention; it is also the only portion of the work that takes an outside look at the nature of the remainder. For the studious "introduction" presents Castalian history no less as a known "fact" than does the Life. Conversely, the motto speaks cautiously of things that are non-existent, not demonstrable, not even probable, but which, by being written about and treated as potentialities, are brought closer to "the possibility of being born."[14]

The "introduction for the layman" appears as a self-contained essay. Less obviously, it is also the basis on which several long passages within the Life build with further elaborations of the themes presented by the fictional chronicler-biographer. As I will show, the introduction has a musical structure that is precisely matched in three self-contained portions within the Life—a fact that seems to have escaped the eyes of Hesse scholars so far.

At the other end of the compound work appear Knecht's own creative products. They, too, are closely intertwined with the Life. The poems are mentioned twice in the story. Once, during Knecht's Waldzell years, the reader learns that young Joseph writes poems as a safety valve for his nascent doubts about Castalia's irreproachability and inviolability. Much later, planning his retreat from his position, the *magister ludi* remembers an additional poem that he composed during his years of Chinese studies.

Knecht's three conjectural biographies are presented as exercises in style, research, and self-knowledge required of all young Castalians during their years of "free study." Having submitted these three "lives," Knecht is gently reprimanded by the Board of Education: were he to write another conjectural biography, he should aim at making it less legendary and anchoring it instead in an historically verifiable era. This "fourth life," though extant only in two extensive but unfinished drafts and therefore commonly not included in the compound work, has since been published separately.[15] I want to argue that its place in the overall plan, as indicated in

[14]Albertus Secundus, as Hesse clarified repeatedly, is fictional, though the name no doubt reflects Hesse's reverence for Albertus Magnus, whom he made a member of the "league" in *Journey to the East*. The alleged editors of the treatise, Clangor and Collofino (Latin and Italian for "noise" and "slender neck"), point to two friends who translated Hesse's German into scholastic Latin, Franz Schall and Josef Feinhals. The title of the treatise is given as "tract. de cristall. spirit." Wim Verbaal, Latinist of Ghent University who specializes in language and literature of the medieval period, reads this as an abbreviation of *Tractatus de cristallorum spiritu*, where the otherwise rare plural of *cristallum*, known primarily from mannered poets like Statius, Claudianus, and Avitus, has the metaphoric connotation of "crystal drops"—a fitting image for glass beads (private communication, March 2004).

[15]English in Hermann Hesse, *Tales of Student Life*, trans. R. Manheim, ed. and with intro. by T. Ziolkowski (New York: Farrar, Straus and Giroux, 1976), pp. 125-194/195-233.

the initial sketch, and its implicit significance in the sequential development of Hesse's ideas make it desirable to consider this "fourth life" a delicately linked yet indispensable further component of the whole.

This "whole," as will have become obvious, is no novel in any normal sense of the word. Hesse shunned the German equivalent and argued against subtitling "Roman" even those of his works that follow a straighter narrative course. In English translation, *Das Glasperlenspiel: Versuch einer Lebensbeschreibung des Magister Ludi Josef Knecht* (quite literally: "attempt at a description of the Magister Ludi Joseph Knecht's life") became *The Glass Bead Game (Magister Ludi): A Novel.* Of course there really is no generic term that would do justice to such an intricately interwoven compound structure. Hesse once admitted that he had dreamed of writing an opera that would represent "life's swaying between the two poles of nature and mind in a serene, playful, and accomplished way," for he regarded opera as "the highest form of the fairy-tale."[16] After giving it much thought he recognized that what he would have liked to compose was Mozart's *Magic Flute*, which already existed; so there was no point in pursuing the idea further. Perhaps it was the unconscious residue of this dream that turned his depiction of an imaginary World of the Mind—contrasted with but indebted to a world of mundane concerns and thus ultimately struggling between aesthetics and social responsibility—into a work that is at once literary and musical.

Playing the Castalian Game: Hints for the Uninitiated

The game around which the roughly 550 pages of Hesse's work revolve is famously elusive. More than one hundred pages into the story, when the protagonist has decided to make a particular game design the basis of several years of free study, the chronicler offers the pithy explanation that a reader who "chanced to be ignorant" of the glass bead game might imagine its pattern as rather similar to that of a chess game, "except that the significance of the pieces and the potentialities of their relationships to one another and their effect upon one another multiplied manifold and an actual content must be ascribed to each piece, each constellation, each chess move, of which this move, configuration, and so on is the symbol" (p. 125).

In a letter Hesse wrote in the early stages of grappling with the "Game" as a symbol, he adds another simile: "Playing the Game can best be compared with playing music. That, too, is not considered a serious occupation for people of world and business affairs, yet for the genuine musician it is the

[16]Translated from Hesse, "Kurzgefasster Lebenslauf," pp. 59-60.

celebration of something positively holy."[17] Six years later Hesse articulated his tight-rope walk between undesirable pragmatism and potentially annoying mystification: "I am gradually approaching the point in my narrative where it would be essential to give a kind of 'concrete' idea of the glass bead game, but without diminishing its mystery. The challenge, as described in the motto, is to treat the non-existent as though it existed. This can only be achieved through magic, in fragmentary visions and incantations."[18]

The information Hesse provides about the two glass bead games with which his hero is exceedingly successful—the one Knecht submits for the annual competition of the Waldzell elite, and the one he composes for his first celebration as *magister ludi*—is teasingly scanty and unsubstantial. On occasion of the former, we read that "the competition involved working out sketches for Games based on three or four prescribed main themes" (pp. 195-196). Knecht wins first prize with a game "in the vein of the old masters," although on other occasions his preference is for "the psychological method of Game construction" where "the object was to create unity and harmony, cosmic roundedness and perfection, not so much in the choice, arrangement, interweaving, association, and contrast of the contents as in the meditation which followed every stage of the Game." As Knecht is quoted as having written, this kind of game "encompasses the player after the completion of meditation as the surface of a sphere encompasses its center, and leaves him with the feeling that he has extracted from the universe of accident and confusion a totally symmetrical and harmonious cosmos, and absorbed it into himself" (p. 197).

This is the spirit we are to imagine as reigning in Knecht's first *ludus solemnis*. The chronicler resorts once again to the chess metaphor when he describes the celebrating *magister* as "quiet, inaccessible, entirely the high priest, white-and-gold-clad major piece on the solemn chess-board of symbols" (pp. 262-3). With its search for spiritual meaning and the aspiration toward a kind of redemption in the final harmony, the game becomes transparent as a secular analog of the Holy Mass: the *magister ludi* is said to use "ritual gestures" and a script considered as a *lingua sacra*; the meditation that concludes each movement or act of the game aims at a kind of mystical-intellectual transcendence, a secular *unio mystica*.

(In Appendix 1, I have compiled six excerpts from Hesse's own pen that shed light on the glass bead game: one from the early sketch, three from observations the chronicler offers at different points in the story, two ascribed to Joseph Knecht, and one representing the Music Master's view.)

[17]Transl. from a letter of mid-June 1932 to Paul Schottky, reprinted in *Materialien* I, p. 56.

[18]Transl. from a letter of 11 January 1938 to Otto Basler; *Materialien* I, p. 183.

Hindemith: Operatic Portrayal of Kepler's Dream of Harmony

Universal Harmony in a Composer's Thinking

For musicians, "harmony" may seem too self-evident a concern—and too specific a concept—to warrant deliberations of a general nature. While any good dictionary of musical terms will list at least two meanings for harmony (variations of "the simultaneous sounding of notes, to produce intervals and chords" and "the successive sounding of chords, to produce progressions"), all connotations imply acoustic features and actual or at least imagined tone production. The philosophical idea of "consonance" as a correspondence of ratios pertaining across various fields of natural phenomena has intrigued only a few composers from the Western tradition.

Paul Hindemith (1895-1963) was well aware of the Pythagorean quest and its foundation in Greek music theory. Both had considerable impact on the theoretical first volume of his treatise *Unterweisung im Tonsatz*, written in 1937. In the English translation, he concludes the introduction with the following assertion:

> [...] I am in agreement with views which were held long before the classic masters. We find such views in early antiquity, and far-sighted composers of the Middle Ages and of modern times hold firmly to them and pass them on. What did tonal materials mean to the ancients? Intervals spoke to them of the first days of the creation of the world: mysterious as Number, of the same stuff as the basic concepts of time and space, the very dimensions of the audible as of the visible world, building stones of the universe, which, in their minds, was constructed in the same proportions as the overtone series, so that measure, music, and the cosmos inseparably merged. [... T]hat our consideration of tonal materials and its application by all who may concern themselves with it may catch a kindling spark from the spirit of the old masters is the hope from which this work springs.[19]

[19]Paul Hindemith, *The Craft of Musical Composition, Volume I: Theory*, trans. Arthur Mendel (New York: Associated Music Publishers, 1945), pp. 12-13.

The Genesis of *Die Harmonie der Welt*

Hindemith first considered an opera about Johannes Kepler and his explorations of universal harmony in 1939. Throughout the war years, he kept mentioning the project in his letters and collected material about persons and issues connected to this topic.[20] Major portions of the music must have existed fully formed in his head before anything—words or notes—ever appeared on paper. In 1951, when Paul Sacher offered to celebrate the 25th anniversary of the Basel Chamber Orchestra with the premiere of a "preview suite" of the future opera, Hindemith responded by quickly writing down and sending off the Symphony *Die Harmonie der Welt* which, in the composer's words, "develops passages from the opera"—a work as yet unwritten. Wilhelm Furtwängler immediately adopted the symphony into his repertoire, declaring that he thought it Hindemith's best orchestral composition. Despite this encouragement, it took the composer another five years until he finally sat down to pen the libretto, which he completed on 1 September 1956. The music then followed swiftly, and the full score was ready by 30 May 1957.[21] The opera was premiered, under the composer's baton, during that year's Munich Opera Festival, on 11 August 1957.

The libretto portrays the protagonist in a way that is utterly conscientious in its attempt both to be true to Kepler's intents and to give a wealth of details. Hindemith sketches the era with its religious bickering and political hazards, its susceptibility to superstition and to adventurism of all kinds. Before this background, Kepler appears as a man of crystal-clear thought, incorruptible volition, and uncompromising character. The picture is of a complex personality; yet Hindemith never makes his own judgment explicit. Perhaps more surprising is the fact that he never hints at the extraordinary

[20]I am grateful to the Hindemith-Institut in Frankfurt and its director, Dr. Giselher Schubert, for his kind permission to study the composer's ample notes, which chronicle his reading and preparatory thoughts on the opera. These notes reveal the impressive breadth of his historiographic research as well as his seriousness to understand Kepler's scientific and religious concerns. One of three note books entirely reserved for research toward this opera contains in its seven sections biographical sketches of members of Kepler's family, portraits of other persons of significance in Kepler's life, drafts for various scenes and text passages, notes on living conditions, frequency of disease and death, scientific discussions, religious controversies, and political developments during Kepler's time, memos documenting Hindemith's studies of issues important for Kepler's professional life, as well as a reading list of more than 60 titles.

[21]For the genesis of *Die Harmonie der Welt*, Sacher's invitation and Furtwängler's praise, see Giselher Schubert, "Unharmonie des Zeitgeschehens: Zu Hindemith's Oper *Die Harmonie der Welt*," in *Hindemith Forum* 6 (Frankfurt: Hindemith-Institut, 2002), pp. 8-11.

impact Kepler and his insights had on the development of science. It remains for the audience to assess the sequence of events and understand (possibly) the lasting greatness and (certainly) the immediate failure of this man.

Judgment may be made even harder by the fact that Hindemith the librettist, mirroring content in form, conceived many of his scenes in the manner of late-medieval woodcuts: in all scenes centering on persons of clear mind, the drama is restricted to the verbal dimension. Only those episodes in which individuals are reigned by irrational forces—mother Kepler's digging in the graveyard for the skull of her long-dead father, Emperor Rudolf's violent distress at seeing the heavens so much more densely populated than he had learned to be comfortable with, and mother Kepler's witch trial—rivet the audience with scenically captured suspense.

Kepler's character, thoughts, and hopes are developed in contrast to historical persons who cross his path or share his life. Foremost among them is the empire's most successful general, Count Wallenstein. His imagination and determination match Kepler's but are directed to opposite ends. While Kepler's goal is to ascertain universal consonance, celebrate its manifestations as proofs of God's glorious world plan, and demonstrate its realization in all mundane realms as humankind's intended state, Wallenstein, as a man of action, is ambitious not to prove but to bring about what he considers "world harmony"—if necessary, by force.

Two other counterparts to Kepler the spiritual seeker and intellectual are his mother, who values her son's learning but scorns its objects, and Hizler, the Lutheran pastor in Linz who defies Kepler's religious plight. Mother Kepler, immersed in occult endeavors and mythical fantasies, lives in a dream world where everything strives toward secret powers. Pastor Hizler, a church functionary who derives his importance entirely from his institution, seizes the opportunity to exercise power by banning a famous scientist and philosopher—a man most likely superior to him in the ability to wrestle with his own conscience—from the consolations of formal worship.

Cosmological and Spiritual Themes in the Operatic Plot

Kepler's life is shown in five stages strung over a period of 22 years: from 1608, the peak of his successful period as imperial mathematician in Prague, to his death in 1630. Act I presents an exposition of the protagonist's person and his professional as well as private concerns:

- Kepler's much-read treatise about the (limited) power of the comet Europe had seen in 1607 is stymied in a market-place speech by Tansur, a university dropout and political agent;

- Kepler's first experiments with the telescope are shared by Emperor Rudolf, whose belief in a comprehensible cosmos is shaken by the sudden profusion of stars and planetary moons;
- Kepler's early horoscope for Wallenstein is evoked when we witness the count asking Kepler's research assistant Ulrich to deliver the request on behalf of "the world's soon-to-be-greatest general";
- Kepler's worries about his mother's notoriety (and eventual arrest for witchcraft) are foreshadowed in a cemetery scene in which she attempts to disinter her father's skull;
- Kepler's love for his children and grief over the recent death of his first wife and small son are expressed in a conversation with his daughter;
- Kepler's preoccupation with the calculation of the planet Mars's orbit is revealed in an exchange with Ulrich.

The act concludes with the people in the street commenting on Emperor Rudolf's forced abdication, the event that ended Kepler's very productive period in Prague.

Besides setting the stage and introducing the main players as well as the protagonist's prominent concerns, the exposition also introduces the most prominent reservations raised against Kepler: his mother fears for her son's soul because of his "intellectual conceit," brother Christoph calls Kepler's endeavors "shadow-boxing," Emperor Rudolf weeps over Kepler's alleged attempt to "destroy the world for me," and Ulrich recoils from "a thinking that respects nothing real."

Act II presents developments of the various themes introduced in Act I. In Prague, Wallenstein's desire to build a palace that will reflect his power and grandeur deprives the city's poor of the slum they have been calling home. In Linz, Kepler's refusal to pretend against his conscience that he has changed his belief about the nature of the Eucharist leads to his exclusion from communion and church attendance, but also to an invitation by baron Starhemberg to consider his foster-child Susanna as a bride and mother for his orphaned children. Ulrich, discontent with the demands of his life as Kepler's assistant and jealous at seeing his master court his childhood friend Susanna, becomes susceptible to the rhetoric with which Tansur seeks to recruit him into Wallenstein's army. The emphasis in this act is on the inner voice and on self-definition. This is played out superficially in Wallenstein's and Ulrich's deliberations on how best to realize their ambitions, and with psychological depth in Kepler's efforts to explain his religious creed to Pastor Hizler and his intellectual and spiritual goals to Susanna.

In act III, Hindemith explores two imaginative developments of the historical Kepler's "Moon Dream" essay. While mother Kepler, having fled from the accusations at home, visits her son's house in Linz, the various family members are shown musing about the Moon, attaching very different significance to its impact on their souls and fates. Back in Württemberg, mother Kepler is brought to court on charges of practicing witchcraft. During the trial, the witnesses, listeners, and prosecutors seem far more superstitious than the accused; even mother Kepler's eventual acquittal, owed to the intervention of her famous son and his connections to higher-ranking officials, relies on "influence" rather than on evidence. Once set free, however, she reveals her true face, disclosing the full extent of her fantastic ideas and grandiose ambitions.

Act IV takes us back to Prague, ten years into the Thirty Years War. On the site from which he had once evicted slum dwellers, Wallenstein has since built his dream palace, in which he is hosting noble guests. Finding Ulrich, now a soldier in his army, peering into the ballroom rather than attending to his duties, he barely restrains his impulse to have him hanged. Kepler, whom the general had summoned, witnesses his former assistant's humiliation and tries to intervene on his behalf, but is soon absorbed in a conversation about the stars and their impact. Wallenstein comments on the revised horoscope he has received and will not listen when Kepler tries to downplay its importance. The general has also read Kepler's *Harmonices mundi* and is excited about its implications. He proposes that it should be implemented—if need be by force—to "create heaven on earth." Kepler has strong misgivings when he understands that, allegedly in order to establish universal harmony, the ambitious count plans to conquer all of Europe and that he, Kepler, is to act as his mathematician, calculating time, weather, people, distances, money, and many other variables of combat. Yet when Wallenstein reveals that he has taken over all debts the emperor still owes Kepler and that he is therefore ideally placed to provide or withhold the means of supply for his family, Kepler is compelled to accept the assignment against his better judgment.

In act V, Susanna is alone in the Keplers' new home in Silesia, worried about her husband, who has recently seemed restless and unhappy. Ulrich, who has come to seek a vengeful encounter with his former master, finds only his childhood friend and reveals to her how cynical and spiteful he has become. In Susanna's eyes, Ulrich's behavior mirrors the enmity she experiences everywhere. Meanwhile, Kepler has arrived in Regensburg, where he has traveled to attend the discussion between the emperor and the electors about their discontent with Wallenstein's endless war. When the emperor gives in to pressure and signs Wallenstein's dismissal, Kepler, sick in bed at a small inn, sees his life once more uprooted as a result of his employer's change of fortune. Suddenly calm and detached, he questions his misguided need to chase after harmony also in professional affairs. He now recognizes that his misunderstanding of the meaning universal harmony should have for his personal life has been almost as profound as Wallenstein's unfortunate military interpretation of the concept. As Kepler feels death approaching, he recognizes that harmony has been accessible to him all along—with his wife and children—had he only had eyes to see it. Now he is approaching the greatest harmony humans can achieve: death. His serenity is disturbed, though, by a local pastor who attempts to cajole and threaten the dying man into abjuring his allegedly wrong ideas about the Eucharist. Kepler laments that "disharmony is yet Death's song of temptation," muses about the weirdness of human beings whose minds and spirits can strive heavenward while their bodies decay and return to dust, and dies in the hope of heaven's harmony.

Hardly has he passed the imaginary threshold than a chorus beckons that he now behold what he once divined: the cosmic harmony. Joining the allegorical transfiguration experienced by the other operatic characters, he turns into planet Earth. Bolstered by the chorus of the Milky Way, the heavenly bodies sing a hymn in which they express a growing understanding of their role in universal harmony.

Part I
MUSIC

PLATE 1: Pythagoras, Boethius, and Confucius
with Variants of the Monochord [1]

[1] "Pitagoras musicus": miniature from Fr. Gaffurius' *Theoria musicae*, 15th-century ms; "Boethius with the monochord": miniature from a Canterbury manuscript, ca. 1150; "Confucius playing the seven-stringed zither": from the woodcut album *Shenji tu*, ca. 1440.

The Paradigm of Embodied Consonance

While numerous aspects of our world exhibit harmonious proportions, human senses discern them most easily in the field of acoustics. When tones are derived from a fundamental through processes of rational partition or extension, even untrained ears perceive the resulting intervals as "consonant". The mathematics that apply here are simple. The fundamental tone created by a vibrating body yields secondary tones according to the factor by which the length of the string or air column is divided or multiplied. This results in vibrations whose speed is inverse to the changed length: a string half as long as another will vibrate twice as fast; an air column five times longer than another will vibrate five times as slowly. In internal correspondence to these external facts, the human ear has a natural propensity to distinguish intervals of simple ratio (1 : 2, 1 : 3, 2 : 3, 3 : 4, etc.) and to perceive simultaneous or successive tones of such proportional values as intervallic entities.

Almost everything we know to this day about the laws governing the consonance of tones goes back to Pythagoras. The philosopher himself was content with defining the ratios of the octave (1:2), the fifth (2:3), the fourth (3:4), and the whole tone (8:9); his followers worked out the ratios for the other natural intervals in the scale. As they thus extended the sequence, their ears told them that not all tones were equally "harmonious." The decreasing distance between subsequent tones eventually yields results that the ear hears as variants of a single interval (two kinds of the major third), but one variant will typically sound more agreeable to innate expectations than others.

In example 1 below,[2] the upper staff shows the *partials*, which are the result of *dividing* the string or air column and thereby increasing the frequency; the lower staff shows the *multiples*, whose vibrating body is several times *longer* while its vibrational speed is accordingly reduced to a fraction. Both series are natural to the sounding bodies with which we make music as well as to our ears' innate sense of harmony, but only the partials occur passively, as overtones alongside their fundamental and thus creating component vibrations in particularly suited instruments, most notably in bells.

[2]I have chosen C as the fundamental for convenience (any note would do), transposed both series to the central range for best legibility, and pragmatically stopped at the 16th partial and multiple. Those tones that, although the result of a fundamental's simple multiplication or division, were often avoided in ancient and common-period scales, appear in brackets.

EXAMPLE 1: The partials and multiples of C

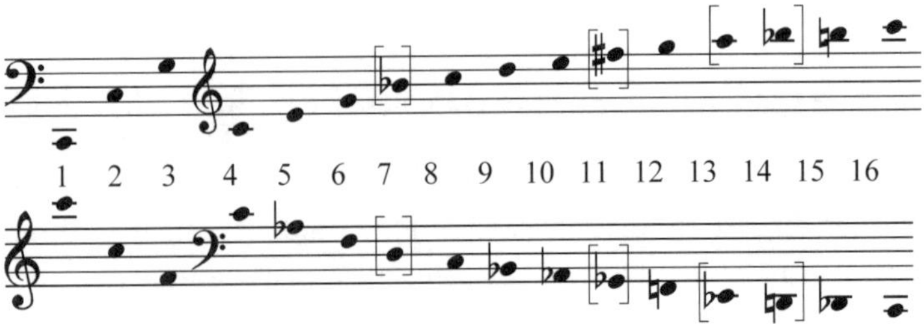

From these two series result the two basic triads of our tonal system: partials 4–5–6 build the major triad, multiples 4–5–6, the minor triad. Just as the second partial or multiple relates to the fundamental as an octave, any other doubling or halving will give equal results (3 : 6 : 12, 5 : 10 etc.).

Only Pythagoras's initial, perfect intervals share two exclusive properties: they first appear as relationships between adjacent tones in the series—the octave at 1 : 2, the fifth at 2 : 3, and the fourth at 3 : 4—and neither the human ear nor the series of natural ratios will allow for any variants. A second set meets with much more lenience from the ear and may have several contenders—this is particularly obvious for the minor third at 5 : 6 and its competitor at 6 : 7 as well as the whole-tone aspirants at 7 : 8, 8 : 9, and 9 : 10.[3] Finally, there is a small group of harmonious intervals built from skipping members of the series of partials or multiples; relevant examples are the major sixth at 3 : 5 and the minor sixth at 5 : 8.

Figure 1 lists, in the order of increasing size, the twelve intervals that occur within an octave and their corresponding ratios.

FIGURE 1: The twelve intervals of the chromatic scale in natural ratios

m2	M2	m3	M3	P4	tritone	P5	m6	M6	m7	M7	P8
15:16	8:9	5:6	4:5	3:4	32:45	2:3	5:8	3:5	5:9	8:15	1:2

The tritone stands out as an interval that occurs naturally only very late in the series, at more than five octaves' remove from the fundamental. This may be one reason why it was long considered suspect—*diabolus in musica*.

Ancient Greek music theory knew three genuses: diatonic, chromatic, and enharmonic. All shared a basic feature: the stepwise progression through

[3]With regard to the three contenders for the whole tone: the modern well-tempered semitone comes close to one half of the ratio 8 : 9, thus endorsing the preference for *do-re*. The fact that partial and multiple 7 have been shunned may originate in number mysticism rather than in reasons of acoustics. Listeners with particularly discerning ears observe that many musicians instinctively intone the seventh partial when playing a dominant-seventh chord, rather than stacking two of the just slightly larger minor thirds at the "accepted size" of ratio 5 : 6.

the octave was divided into two tetrachords, which were framed by perfect fourths and linked by a whole-tone step. This fixed tonal skeleton seems to have been reproduced on at least one of their favorite instruments: scholars of ancient harmonic thinking believe that the lyre known to the Pythagoreans must have had eight strings, and that four of them were fixed in their tuning at the ratio

$$\mathbf{6 : 8 : 9 : 12}$$

while the intermediate ones could be adjusted to accommodate the chosen genus. In this tuning—the ideal of harmonic thinking, as I will show later—6 : 12 equals 1 : 2, the ratio underlying octaves; 6 : 9 and 8 : 12 are analogs of 2 : 3, the perfect fifth; 6 : 8 and 9 : 12 correspond to 3 : 4, the perfect fourth; and 8 : 9 is the preferred version of the whole tone. The foursome thus comprises, and is perfectly limited to, Pythagoras's four basic ratios.

The major systematic extrapolation of the Pythagorean insight into the ratios forming partials and multiples of a fundamental is the *lambdoma* matrix, also known as "the Pythagorean table" or "the *abacus*." The term *abacus* in this meaning goes back to Boethius's *De Musica*, Book II: 19, where he introduces the simple, two-limbed version. The word was later applied to the calculating instrument with stringed beads that travelers to China saw East Asian merchants use. Originally, however, it was much more than the simple calculating tool as which it appears at first glance. The Pythagoreans believed that the *lambdoma* matrix encoded wisdom about many fields beyond music, including speculative arithmetic, geometric forms, planetary orbits, colors, architectural proportions, room resonances, and even an understanding of the divine conception of the World-Soul.

The skeleton of the *lambdoma* matrix consists of the two series of partials and multiples, arranged in the form of the Greek letter *lambda*:

$$\Lambda$$

The space between the two limbs is then filled logically, i.e., with transpositions of the two basic numerical series. This may sound simple to a modern mind, yet the table was considered secret in the Pythagorean brotherhood. In the early 4th century C.E., Iamblichus mentions the matrix in his comment on Nicomachos of Gerasa's arithmetic of ca. 140 C.E., apparently without being aware of its full meaning. In the Renaissance, Salinas and Zarlino go no further. After epochs of virtual oblivion, the *lambdoma* matrix was only rediscovered (and, in part, reconstructed) by Albert von Thimus in the 19th century.

FIGURE 2: The *lambdoma* matrix, as reconstructed by Albert von Thimus

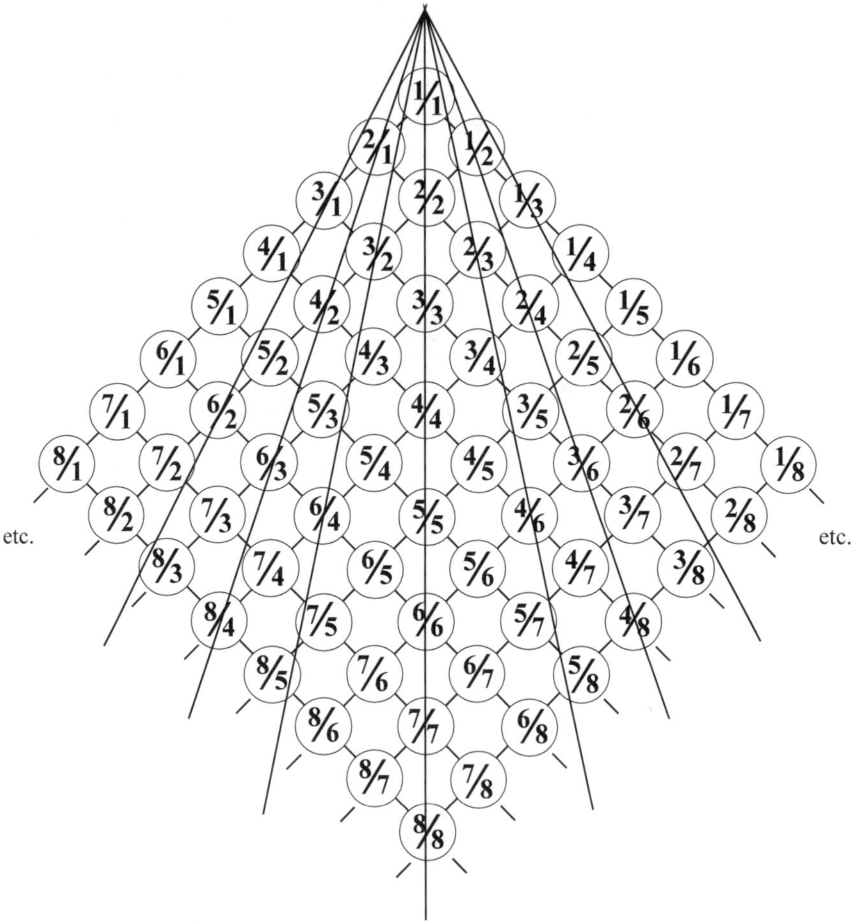

In the interest of uniformity, Thimus deviates from the whole numbers used in one limb of Boethius's drawing and instead presents all values of the two series as fractions. The radiating lines all run through fractions that are different instantiations of the same value (hence, in music, the same tone), with the central vertical linking alternative expressions of the fundamental. All these lines have their source in a point "beyond" the fundamental. This fact, which seems irrelevant for the matrix's application in music, was read by those interested in its spiritual significance as pointing to a creator god outside the material world as the ultimate source.

Finally, the diagram reveals a very basic feature: all fractions on a diagonal share either the numerator or the denominator. A striking analogy of this structure exists in a table from the Chinese *I Ching*. This ancient "Book of

Changes," to which I will return in a later chapter of this study, consists of sixty-four commented hexagrams, each composed of six lines that are either broken or unbroken. The complete set of hexagrams was often represented in tables. Rudolf Haase[4] reproduces one that is attributed to Fu Hsi, the legendary cultural hero and sage credited with the invention of the trigrams.[5] The combination of the two trigrams that make up each sign shows all the features just ascertained for the *lambdoma* matrix; see particularly the identity of the upper or lower trigrams in the diagonals and the analogy of the hexagrams on the central vertical. Haase interprets this finding as proof that the Pythagorean table is a symbol for a secret knowledge discovered independently in several ancient cultures.

FIGURE 3: The *I Ching* in Fu Hsi's matrix

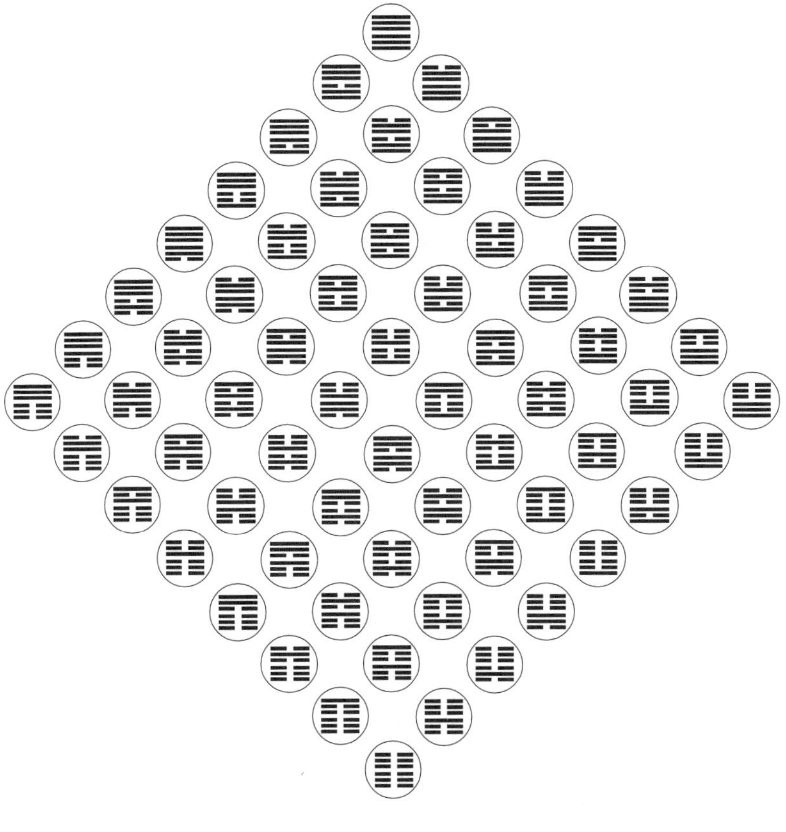

[4]Haase is the most prolific late-20th-century scholar in the field of harmonic research.

[5]Rudolf Haase, *Harmonikale Synthese* (Vienna: Lafite, 1980), p. 68-69. The original Chinese table looked different since it would have been read, like Chinese texts, from the lower right to the upper left. Haase's rendering rearranges the sequence for Western reading habits.

One final item in the discussion of harmonic proportions in music must be mentioned: the instruments the Pythagoreans used to check their theories and demonstrate their findings. Three are relevant here: the lyre, the monochord, and the helicon.

- The lyre with its four strings tuned to the ratios 6 : 8 : 9 : 12 and four intermediate strings variously adjusted to the desired genus has already been mentioned. (The *kithara* was similarly built.)
- The monochord consists of a sounding box over which a single string is stretched. Its moveable bridge slides on a kind of ruler that allows to measure the length of a portion of the string and thus calculate its ratio to the remainder or the whole. The monochord was certainly far more significant for scientific inquiry than for musical performance. Additionally, Pythagoreans mention it as a training tool for singers.
- Helicon, originally the name of a large mountain in Boeotia (a sacred site allegedly favored by gods and goddesses; Greek mythology and poetry linked particularly the muses with Mount Helicon), became the name of two different kinds of musical instruments. The modern brass instrument (a kind of tuba or sousaphone often used in march music) does not interest here. The helicon the Pythagoreans knew is a string instrument used, once again, for reference in music theory.[6] It consisted of a sound box whose surface, unlike that of the monochord, was not long and slim but square. It had a unique design:

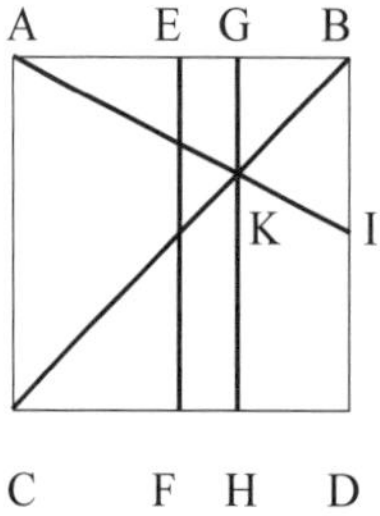

In the square ABCD, sides are halved at E and I.
The diagonal B–C cuts AI at K.
A vertical through K creates G–H.
4 strings of 3 lengths (E–F/G–H, A–I, B–C) are each intersected at the octave, the fifth, and the fourth.
The verticals too, seen from right to left, are spaced at the ratios 6 : 8 : 9 : 12.

Lyre, monochord, and helicon were all designed for one major purpose: to sound the basic values of the secret scheme of numerical ratios. In music, the cell of any harmonic experience is not the single tone but the interval. The Pythagoreans were convinced that the ratios they had discovered for the principal musical consonances also govern each aspect of the universe.

[6]Mentioned by Aristides Quintilianus and Claudius Ptolemy; see Rudolf Schäfke, ed., *Aristides Quintilianus: Von der Musik* (Berlin: Hesse, 1937), p.313; I. Dühringer, "Ptolemaios und Porphyrios über die Musik," in *Göteborgs Högskolas Årsskrift* 40/1 (1934): 61.

Archetypal Harmonies in Hindemith's Symphony and Opera

In addition to being a realm of sounds that is directly accessible to the ear and a frame for non-verbal references, music itself can also be a symbol. When we say that something a person says "is music to our ears," we are usually not talking about the timbral qualities of the person's voice, but of the way in which the content expressed is consonant with our hopes or needs. As Boethius taught the world in the early 6th century, this symbolic notion of music comprises three kinds: the music made by human instrumentalists and singers, the harmony of the human body and soul, and the consonance reigning in all manifestations of the divinely created world.[7]

When Hindemith composed his *Sinfonie "Die Harmonie der Welt"* purportedly from excerpts of his as yet unwritten opera, he organized his material into three movements. Their German indications for tempo and character, which one can translate to the more familiar Italian terms as *Largo – Marcia, pesante – Vivace strepitoso* for the first, *Molto sostenuto – Tranquillo (like a wistful dance heard from far away)* for the second, and *Largo molto – Passacaglia, tranquillo – Agitato* for the third movement, surprise with their unusual prevalence of slow and reflective moods. More intriguing is the fact that Hindemith entitled the three movements with the terms for Boethius's three kinds of music. What the 1952 listeners necessarily heard as absolute music (since the opera to which they might trace the thematic material would not exist until five years later) was thus determined in the composer's mind as indicative not primarily, or at least not exclusively, of specific dramatic events, but of the different ways in which humankind can experience universal consonance—and thus of the future opera's spiritual content.

The composer's brief introductory comment, published in the program book prepared for the Basel premiere, is useful particularly with regard to the general theme: it informs the symphony audience that the prospected opera will portray Kepler "and his search for the harmony that undoubtedly governs the universe,"[8] that the titles of the symphonic movements "refer to a classification often found in the ancients," and that the composer chose these terms so that they may "point to all those earlier attempts at recognizing universal harmony and understanding music as its sounding parable."

[7]The latter is Boethius's *musica mundana*. In Latin, *mundanus*, "referring to the world," is a value-free adjective and, unlike the English "mundane," not an antonym of sublime.

[8]This and subsequent passages are translated from Walter Gerstenberg, "Hindemiths Oper *Die Harmonie der Welt*: Historische und musikalische Perspektiven," in *Programmheft zur Uraufführung* (Munich: Prinzregententheater, 1957), pp. 34-46 [37-38].

When it comes to specifics, Hindemith's designations of the final movement as "trying to symbolize the postulated harmony of the world in a musical form" and of the central piece as "taken from scenes that deal with matters of relevance for the dramatic characters' souls" are helpful. Less satisfying, and more in need of further investigation, is his description of the first movement as "containing music from operatic scenes in which adverse external circumstances hamper the protagonist's actions."

Now that the operatic plot as well as the full score are available and we can ascertain from which of the operatic scenes the musical passages derive, it is intriguing to observe what events Hindemith in this work connected with Boethius's three terms. The *musica instrumentalis* movement begins with what will turn into the operatic overture. This is followed by excerpts from the March that opens act II. In the operatic plot, this scene has Tansur, on behalf of Wallenstein who later enters to second his employee, chase slum dwellers from the site on which the prosperous general intends to have his palace built. The third component corresponds to the refrain and first episode in the witch-trial scene, featuring the magistrate's reading of the accusations. In all these passages, Kepler is conspicuously absent, both from the dramatic point of view and with regard to the symbolic significance of the musical material. The protagonist will not be seen on stage until the middle of act I and thus long after the overture has ended; he does not have any direct contact with Wallenstein until the beginning of act IV, and none whatsoever with Tansur; and he will not enter the court room in Württemberg until after the refrain material has last sounded in its original format. Rather than to "adverse external circumstances that hamper the protagonist's actions," the music in the two dramatically realized components points to situations of desperation experienced by the most vulnerable among Kepler's contemporaries: those who are too poor to have proper homes, and those whose thoughts and beliefs differ enough from the majority as to threaten their good standing within society and jeopardize their protection from wilful incrimination. But then again: In the course of his life, the historical Kepler himself experienced unemployment and with it, the threat of poverty, as well as punishment for deviant thinking. In this larger and indirectly symbolic sense, the music of the *musica instrumentalis* movement does indeed refer to paradigmatic instances in which the protagonist's search for harmony is adversely affected.

The music heard in the second movement stems from what will years later turn out to be act II of the opera. 58 measures in very slow tempo prefigure the scene in which Kepler, speaking to his future wife, expresses his gratitude for her confidence and articulates his faith in

> new research into the most profound order and magic of the laws as manifested in music, geometry, in the structure of plant and animal, the course of the stars, in essence, motion, relation, in all that has been created. Which also includes the place, thought, deed, and destiny of human beings; from which one should learn to deduce the divine Fiat, the magic Word—spoken at the beginning by the Creator—that brought forth the world and preserves it in Light. Should I not exhort the thinkers, artists, rulers of this earth to accommodate themselves within the confines of their work to that Word, so that through them, humankind will become conscious of the harmony of the world.[9]

The symphonic movement continues with what in the opera will become the subsequent conversation between Kepler and Susanna about their shared creed and the basis of their marriage. Susanna interprets universal harmony for herself as "with the best that resides within us to be closer to our neighbor, understand him, better the evil, forgive, destroy hostility, propagate love." She believes that the net effect of such striving for harmony would be to "root out all weaknesses, serve what is noble, the ideal of man's being, an image of heaven."[10]

The short contrasting passage with its "wistful dance" (cf. from [16 D]) establishes a very soft, languishing $\frac{3}{4}$ time. Before this backdrop, the solo violin plays a high-register echo of Kepler's creed motif in *pp dolcissimo*. The movement concludes with 28 measures that were to end act II. For the opera, Hindemith superimposed over this passage a vocal line in which the young bride articulates a stunning outlook on her marriage.

> I stepped into an Eden
> of numbers, calculations, proportions
> where formulas live instead of snakes,
> apples are harmless spheres.
> No flaming sword will drive us out;
> I will be happy to stay with Adam
> in his wisdom and stillness.
> Blessed am I, blessed am I.

The symphony's third movement corresponds to the music for the final segment of act V (cue [28] in full); Hindemith later merely added vocal lines to the orchestral texture, as well as interpolating a few recitativic inserts. The

[9]See Paul Hindemith, *Die Harmonie der Welt. Textbuch* (Mainz: B. Schott's Söhne, 1957), p. 35; facsimile full score, act II, cue [15].

[10]For this main section of the symphony's second movement, cf. cues [15]-[15 F] in the operatic score.

movement thus prefigures the gathering of the eight luminous allegories in the starry sky and their praise of true harmony. I will discuss this music in detail in a later chapter.

While a quick glance at the symphony may suggest that its movements translate the ancient view of the realm in which harmony ideally prevails into a juxtaposition of human experiences of disharmony (I), imperfect harmony (II), and blessed harmony (III), there is one crucial component that offsets this interpretation: the overture. Its 33 measures are firmly rooted in E. Throughout the first 25 measures, E sounds as a pedal in the timpani; in mm.1-9 it is reinforced by violins and violas. Approaching these pedals, the strings ascend in scalar runs whose pitches, E F G A B C D E, establish the tonal center and the mode on which the overture will be based. In the world of Pythagorean thinking to which the operatic title alludes, this was *the* fundamental scale, called "Dorian." (Not until the Middle Ages, when the Church reassigned the Greek terms, did the mode on E become known as "Phrygian.") The way in which Hindemith opens his work can be interpreted as a deliberate nod to the overarching subject matter. He does not, however, quote the scale on which ancient Greece based its music theory in its neutral form, which would have been descending, slow, and serene. Instead he fashions an emotionally gripping upward burst in septuplet 32nd-notes.

In measure 2, the trumpet introduces the overture's one and only motif. It consists of a mordent on E followed by a descent through six fourths. This descent, broken by an octave reallocation at the point of enharmonic change, thus links the tonic E to its tritone B♭. The plunge is complemented by a slow scalar ascent launched from the note F, the next step in the circle of fifths.

EXAMPLE 2: The all-determining motif in the overture

The pitches of this motif are symbolic in two ways. The mordent E-F-E employs the tones that, in Kepler's famous calculations, correspond to the two tones of earth's orbit, MI and FA. As Kepler interprets this in a marginal gloss toward the end of chapter 6 in Book V of *Harmonices mundi*, "The Earth sings MI FA MI, so that even from the syllables you may guess that in this home of ours, MIsery and FAmine hold sway."[11]

The remainder of the motif then travels clockwise through the circle of fifths, skipping only twice, at the very end, over intermediate tones:

[11]Kepler, *The Harmony of the World*, p. 440. Kepler, writing in Latin, uses the "fixed DO" still common today in Romance-language speaking countries, not the movable DO of solmization.

FIGURE 4: The "order" in motif 1

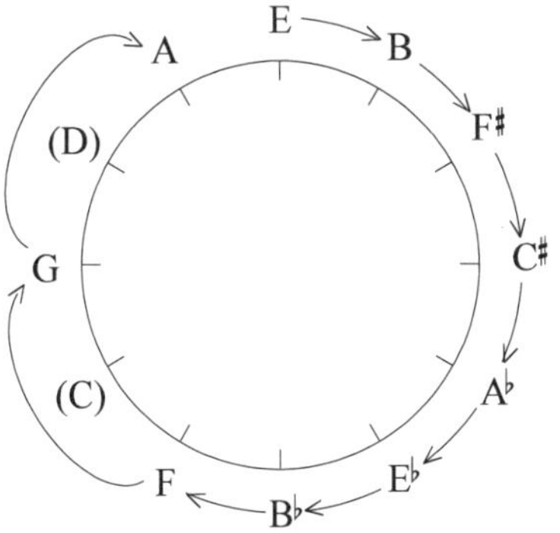

In the course of the overture, motif 1 is heard 12 times in unison on E; an additional entry sounds in 4-part parallel.[12] The 12 regular entries fall into 4 + 5 + 3 (mm. 1-9, 12-21, 24-31). Hindemith uses the numbers for a play with metric ratios. Thus the motif consists of 9 ambiguously grouped beats: while the bar-line divides it into 4 + 5 quarter-notes as shown in the example above, listeners tend to hear the weighty half-note as a down-beat, and the motif therefore as welded of 5 + 4. Moreover, the entire overture presents an irregular alternation of 16 measures in 4/4 and 16 in 5/4 time—(with a single 2/4 exception in m. 27 as part of an extension). For listeners' ears, this "regular irregularity" is as puzzling as are the planetary motions in the starry sky to beholders' eyes. Also, the polyrhythmic pattern that the high strings establish as a backdrop for the overture motif and maintain throughout the first nine measures, presents a juxtaposition of 4:3 divisions in every eighth-note beat. This is more than the Ravelian sparkling organ point: when the pattern gives way, it is substituted by an even more intricate one, a play with tone-pairs in which 3s hold their ground within a notated 4/4 time. The fourth horn repeats the descending step A♭-G as ♩‿𝅗𝅥 ♩‿𝅗𝅥 ♩‿𝅗𝅥 while the low strings play E-D♯ and C♯-B♯ respectively, both in the rhythm ♩.‿♩. ♩.‿♩. ♩.‿♩. but in a stretto at an eighth-note's distance. The three-layered repeated descending step sounds entirely outside the metric framework.

In terms of its structure, rhythm, and meter, the overture thus welcomes us into a world governed by harmonious yet all but simple proportions.

Tonal Patterns in Layout and Thematic Material

Hindemith began thinking about the idea of a single principle ordering the entire universe in the early 1930s, when he conceived the first volume of his textbook *The Craft of Musical Composition*. Several Hindemith scholars

[12] [1] mm. 1-2 trps, [2] 3-4 hrns, [3] 6-7 trbs, [4] 8-9 bsns/trbs/tba; 10-11 new beginning; [5] 12-13 fls, [6] 14-15 obs, [7] 16-17 cls, [8] 18-19 strgs, [9] 20-21 bcl/bsn/hrns/vc;
— 22-23 obs/bcl/bsn/hrns/trps/vc/db on E/A/E/C/F♯ —
[10] 24-25 fl/cl (without ascent), [11] 28-29 bcl, [12] 30-31 bsn/cbsn/db
(26-27 extension, 32-33 liquidation with motivic snippet).

have argued that his contact with Hans Kayser, a student of the Pythagorean quest for universal harmony, may have been instrumental in shaping Hindemith's theory about tonal relationships.[13] Especially the first of his two "series," which is determined by the graded relationship of each tone to a reigning fundamental, invites comparison with the solar system. The planets, orbiting in speeds related to their distance from the sun, compare to the ever remoter tones or tonal centers that Hindemith envisaged circling around their tonic in correspondence to ever weaker degrees of consonance.

FIGURE 5: Hindemith's "series 1," a cosmic tonal order, shown here on E, the main note of Pythagorean harmonic thinking and the tonic of his opera

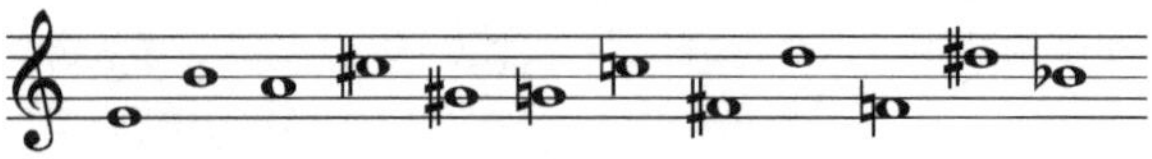

For the cover of his piano work *Ludus tonalis*,[14] Hindemith created a drawing that arranges the twelve tones of "series 1" in a kind of cosmic constellation. Here is a similar "orbital arrangement" around the fundamental E:

FIGURE 6: Hindemith's "series 1," in planetary arrangement around E, the fundamental of *Die Harmonie der Welt*

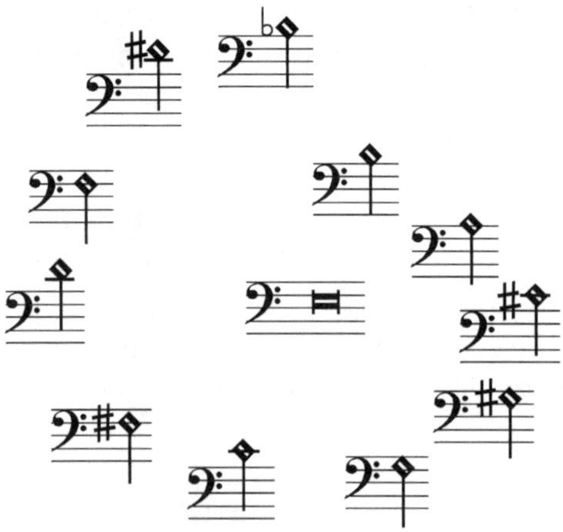

The language the composer used when commenting on his concept in the first volume of *The Craft of Musical Composition* also corroborates the genealogy of his tonal thinking:

[13]See the documentation of his correspondence and encounter with Hans Kayser in Rudolf Haase, "Paul Hindemiths harmonikale Quellen: Sein Briefwechsel mit Hans Kayser," in *Beiträge zur harmonikalen Grundlagenforschung* (Vienna: Lafite, 1973).

[14]Paul Hindemith, *Ludus tonalis* (New York: Associated Music Publishers, 1943).

> Series 1 shows us the distance of the planets from the central star. As the distance increases, the warmth, light and power of the sun diminish, and the tones lose their closeness of relationship..[15]

Surprisingly, attempts at tracing the theory of "decreasing light and warmth" in the very tonal layout of *Die Harmonie der Welt*—a work that would seem a likely candidate for its application—remain largely unsuccessful. While both the overture and the cosmological transfiguration at the end of act V are solidly rooted in E, I find no conclusive evidence that supportive characters and situations are linked to closely related tones (such as the fifth, B, and the fourth, A); nor are the remotest tones reserved for the most inimical persons or circumstances.[16] (In part 4 of this study I will explore a different, metaphorical rather than tonal reading of the planetary scheme, and show that in this sense, planetary order does play a major role in the opera.)

A look at the two sides of the latter aspect—the tonality of adverse components and the use of the tritone B♭—may corroborate my claim. None of the scenes in which the operatic protagonist reaches low points with regard to public recognition or self-esteem emphasizes the tone B♭. Pastor Hizler's damnation of Kepler's alleged heresy in act II (cf. [11 C5-I3]) is rooted in E's close relative A. The scene at the end of act IV, in which Kepler reaches the nadir of his self-negation as he consents to be recruited by Wallenstein despite his reservations, begins its strophes centered in E♭ ([24] and 24 D]). While Kepler, though flattered by Wallenstein's interest, is still skeptical about the application of his concept to military scheming, the music briefly touches B♭ ([24 B and K]). But when Kepler caves in (at [24 F6]), and again when he unhappily tries to convince himself that combining the theoretical idea with the practical application is something he has long wished to explore (at [24 M7]), the music turns to the "supportive" tones A and B. The final trio, in which Tansur's presence confirms how ill-advised Kepler's association with Wallenstein is, even emphasizes A and E. Similarly, Tansur's treatment of Prague's paupers in the opening scene of act II, arguably among the most pitiful events in the opera, is moored in E and thus seems to mock

[15]Paul Hindemith, *The Craft of Musical Composition, Book I,* p. 57.

[16]I remain unconvinced by James d'Angelo's essay on the subject, "Tonality Symbolism in Hindemith's Opera *Die Harmonie der Welt*," published in volume 1985/XIV of the Hindemith-Jahrbuch. The superficial hindrance to the reader, who must figure out that all six headings on pp. 106-117 become meaningful only if one ignores the final accidental, is only a minor annoyance. More crucially, d'Angelo's perceptions of local tonality, often grounded in no more than individually highlighted tones, and of regional tonality based on an act's final chord, come across as largely personal preferences. On this basis, his argument about the associative meaning in pairs of what he calls "tritone-away tonalities" is not persuasive.

this tone's association with Kepler's quest for universal harmony. By contrast, the allegedly adversarial tritone B♭ is prominent in instances that show Kepler puzzled but compassionate: during Emperor Rudolf's raving aria in act I (at [4-4 C]) and at the very end of act III, when mother Kepler is crestfallen after her son's rejection of her plans.

I thus want to argue that by 1957, Hindemith had abandoned the idea of attaching symbolic implications to tonal levels, an idea first expressed in the 1930s and defended extensively in the preface of his revised *Marienleben*.[17] In *Die Harmonie der Welt*, Hindemith seems to have thought along different lines. One of them—the only one I will pursue here so as not to overstate this point—is a thinking in terms of chromatic neighbors rubbing shoulders, as can be found in act I. Tansur's framing song is rooted in F♯. This note is separated from the E of Kepler's quest for universal harmony by different passages rooted in F. When Wallenstein recognizes Tansur's advertising genius and invites him to become his military recruiter (in [2]), he sings with a constant grounding in F. Mother Kepler's eerie cemetery song (in [3]), while accompanied with a dissonant chord comprising C, D, E♭, F and F♯, is melodically moored on F. Emperor Rudolf's first small aria (in [4 B-C]), suspended over repeated accompaniment figures in B♭ minor, frames his roaming through F♯ major/minor and other chords with ever-recurring Fs; in his second small aria (in [6 A]), which marks his return to calm after his distraught outbreak, he regains his mooring on F. Kepler's conversation with his daughter (in [7]) uses the same grounding, as does his song of mourning. And while his interchange with Ulrich, launched from a dissonant chord over F, subsequently digresses to other tonal realms, Kepler returns to F as his melodic frame (in [8 F]) when he asserts that he intends to heed his duties.

Even more than momentary tonal centers, Hindemith's use of melodic intervals in the thematic material is designed to typify the *dramatis personae* or the nature of dramatic situations. To stay with examples from act I: Tansur's framing song, whose lively compound triple time and fairground-style accompaniment hint at his crowd-pleasing cynicism, features a melody of swinging rhythm and memorable contour. Wallenstein's brief but much-repeated vocal phrase with its strong tonality and catchy rhythm shows him as a self-assured, charismatic character and thus oddly close to the wily Tansur. Both shrewd men thus suggest a (musical) proximity to persons without guile, like Kepler's daughter, whose attempt at consoling her father turns on two reiterated vocal phrases of folktune-like simplicity.

[17]For a thorough discussion both of Hindemith's tonal views as laid down in this text and of the discrepancies between stated rationales and actual facts, see my study *Musical Ekphrasis in Hindemith's* Marienleben (Amsterdam/Atlanta: Rodopi, 2000), particularly pp. 76-79.

EXAMPLE 3: Tansur's publicity song [1]

EXAMPLE 4: Wallenstein's practical wisdom [2]

EXAMPLE 5: Kepler's daughter sharing her father's grief [7]

The opera's two troubled figures have very different music. Mother Kepler's grave-digging song is accompanied with muffled woodwind chords replete with trills and percussive rolls. The reticence of the halting pulse matches the creepiness of her ambition: to defeat her eldest son's "intellectual conceit" by making him drink from a goblet fashioned from her dead father's skull. Her vocal line rises in several ascents—octatonic, chromatic, and hybrid—and concludes in a 7-tone octave descent of alternating minor thirds and minor seconds, which thereafter echoes throughout the opera as a tonal symbol for the proximity of piety and superstition.

EXAMPLE 6: Katharina's song of madness and sorcery [3]

Emperor Rudolf's musings about the unreal world in which he lives and feels lost, and the consolation he finds as he immerses himself in the sight of the ordered cosmos, is clad in a 7-tone octave ascent that mirrors mother Kepler's descent. Rising intervals of increasing size (a major third, a fourth, and a fifth), each chromatically bent back downward, and a rhythm studded with syncopations speak of his enthusiasm, but also of his insecurity and the shakiness of his sense of reality.

EXAMPLE 7: Emperor Rudolf's longing for cosmic solace [4]

Wenn die Schein-welt meines Herr-scher-seins
(When the phantom world of my sovereign-existence)

As the emperor gazes at the star-lit sky through Kepler's new telescope, he gasps at the sight of a cosmos "swarming without aim or end, a pail of glistening refuge tipped out by a madman," since he expects the heavenly order to prefigure what should be possible on earth. The music expressing his distress in [5] is based on a jagged motif in which two plunging major sevenths are linked by an ascending fifth. The rhythm of this very lively motif is a portrayal of the sovereign's state of mind: he holds fast to one aspect, only to abandon it abruptly for another, in a mental course in which despair prevails over briefly flaring hope. Initially restricted to the orchestra, the motif eventually even takes hold of the emperor's vocal line (at [5 F]).

EXAMPLE 8: Emperor Rudolf's agitation [5]

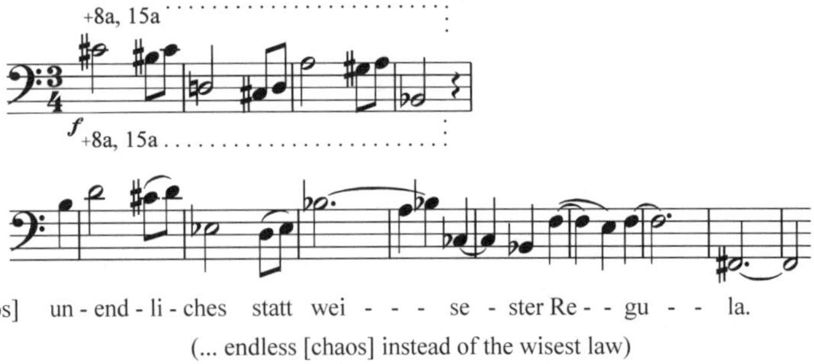

[Chaos] un - end - li - ches statt wei - - - se - ster Re - - gu - - la.
(... endless [chaos] instead of the wisest law)

This kind of musical signification is neither Hindemith's invention nor is it reserved for this opera on universal harmony. It is a way of creating non-verbal reference that is widely understood by listeners at home in the context of Western classical music, and thus all the more apt to convey messages.

Music's Moral Power in Ancient China and Hesse's Castalia

Music is the most discernible force in Hesse's *Glass Bead Game*. Some of its workings are obvious: the game is said to have been invented by musicologists, a main step toward its universal elaboration was the achievement of "a Swiss musicologist with a passion for mathematics" (p. 36),[18] and the designs of the elaborate games performed and celebrated in Castalia are typically described in musical terms. Joseph Knecht starts out as a Latin school student particularly notable for his musical talent, his admittance into the system of elite schools is decided in a test of chamber music playing, his introduction to meditation is based on music, and he spends his first years in Waldzell focusing on music studies. His life-long mentor is the Music Master, his oldest friend, Carlo Ferromonte, becomes a musicologist, and his emotional connection with his last mentor, Father Jacobus, rests on their shared love for the music of the early 18th century. Extensive discussions of musical styles and structures, musical experiences, and music's role in the life of the mind add up to what constitutes the largest component of the reflective material interspersed with the narrative portions of the biography.

Perhaps most importantly, both the protagonist and his biographer recognize music as a power that is at once esthetic and ethical. Especially insofar as it is a fundamental component of the glass bead game, music is valued in this dual role. At times, the two aspects are assigned to Western classical and ancient Chinese music respectively, but ultimately, only that music in each tradition which combines them is considered worth striving for. In Knecht's words, "in the final analysis every important cultural gesture comes down to a morality, a model for human behavior concentrated in a gesture" (p. 43).

Music arises from the harmony of thinking and feeling, the two poles of life. If either pole–intellectuality or sensuality–prevails, then music degenerates, marking a decline in the civilization that produces it. Knecht regards the music of the 19th and 20th centuries as degenerate since in this period, the sensuous elements outweigh the intellectual. But the opposite is equally dangerous: he reminds his glass bead game students that music is made with hands and fingers, mouth and lungs, not with the brain alone, and that it will not do to abstract the purely intellectual from the full substance.

Three extensive references to ancient Chinese music endorse just how significant music's moral effect must be understood to be. The chronicler in his introduction includes a long quotation that lays out the Confucian teaching of how music is linked to and mirrors socio-political situations. It begins with the significant tenet: "The origins of music lie far back in the past.

[18]All page numbers refer to the 2002 Picador edition of *The Glass Bead Game*.

Music arises from Measure and is rooted in the great Oneness" (p. 29). The deliberations are excerpted from a book compiled between 265 and 239 B.C.E. at the request of Lü Buwei, regent of Qin during the First Emperor's childhood.[19] The chronicler prefaces them with the general observation that "in the legendary China of the Old Kings, music was accorded a dominant place in state and court. It was held that if music throve, all was well with culture and morality and with the kingdom itself" (p. 28). Wedged between these more philosophical observations is a passage that hints at dramatic anecdotal evidence.

> [...] The poets told horrific fables about the forbidden, diabolic, heaven-offending keys [...], the "music of decline"; no sooner were these wicked notes struck in the Royal Palace than the sky darkened, the walls trembled and collapsed, and kingdom and sovereign went to their doom (pp. 28-29).

The allusion points to an episode included in a Chinese historical novel about the decline of the Zhou Dynasty (1066-256 B.C.E.).[20] Hesse had told a fuller version of this "fable" in his 1920 novella, *Klingsors letzter Sommer* [Klingsor's last summer]. The tale features a princely ruler who brings about his own downfall and that of his people and state by insisting on listening to music composed in a fateful mode—music that a visiting sovereign's court musician had overheard and delightedly memorized *en route* but which the prince's own blind court musician warns him is pernicious.

The third reference to ancient Chinese music and its moral value ties it back to the Castalian story's narrative cast. During his years of free study, Knecht devotes several years to an exploration of this field. As his biographer tells it,

> Everywhere in the older Chinese writers he encountered praise of music as one of the primal sources of all order, morality, beauty, and health. This broad, ethical view of music was familiar to him from of old, for the Music Master could be regarded as the very embodiment of it (p. 133).

The Music Master, whose field is Western music but whose portrayal throughout the Life is strongly reminiscent of the image of a Chinese sage,

[19]For a recent American translation see John Knoblock and Jeffrey Riegel, *The Annals of Lü Buwei* (Stanford, CA: Stanford University Press, 2000). The *Annals* consist of 160 chapters, combining the teachings of Chinese philosophers from the 12th to the 3rd centuries B.C.E. The excerpts Hesse quotes stem from Book 5, Chapters 2-4 (cf. Appendix II).

[20]Hesse knew the fable from a book he had reviewed, Leo Greener's *Chinesische Abende: Märchen und Geschichen aus dem alten China* (Frankfurt: Insel, 1913).

is repeatedly described as "serene." This adjective seems like a code word for harmony in both humans and music. In the excerpt from Lü Buwei we read:

> Decaying states and men ripe for doom do not, of course, lack music either, but their music is not serene. [...] What all sacred sovereigns have loved in music was its serenity. The tyrants Giae and Jou Sin made tempestuous music. They thought loud sounds beautiful and massed effects interesting. They strove for new and rare tonal effects [...] Because it has departed from the essence of real music, this music is not serene. If music is not serene, the people grumble and life is deranged (pp. 29-30).

And Joseph Knecht is quoted as teaching:

> Classical music as gesture signifies knowledge of the tragedy of the human condition, affirmation of human destiny, courage, cheerful serenity (p. 44).

The link between "music" and "serenity" is itself a Chinese one: both are written with the same ideogram (although the pronunciation is unrelated).

[yuè] music	樂	[lè] serene, or cheerful

Musical Structures in the Tale of the Rise of Castalia

Numerous Hesse scholars have commented on the content, purpose, and tone of the chronicler's preface—entitled "The glass bead game: a general introduction to its history for the layman." The chronicler's pedantic pose and his copious deviations from his alleged topic mislead many readers to conclude that he must be "rambling." Yet this introduction has in fact a most interesting and carefully designed layout. As a musicologist addressing a literary text, I have submitted the 33-page passage (pp.11-44 in the Picador edition) to a musical analysis: stating not only the appearance of themes and motifs, formulas and cadenzas, but recording their frequency, distribution, length, and interrelatedness.

The introduction falls into five sections. Sections 1, 2, 4, and 5 are of almost equal length, while the central section 3 is slightly shorter. The first section opens with some preliminary remarks addressing the chronicler's intent and justifying his undertaking: to present to his 25th-century readers the biography of a particular glass bead game master from the early 23rd century, followed by that man's unpublished writings and prefaced with reflections about the distant age that immediately preceded the intellectual reformation in the late 20th century. The preliminary remarks are briefly taken up and complemented by closing remarks at the end of the fifth section. This rudimentary symmetry frames a "composition" that turns on two themes and three motifs. Its predominant recurring "formula" is the disclaimer; as climactic cadenzas we hear three substantial quotations.

Section 1 (ending on p. 18 with "the most important stages in that history") presents both themes, offering first thoughts about the strange era of decay out of which the new order grew as well as a brief glance at the new order's paradigmatic symbol, the glass bead game. Three times does the chronicler begin his preliminary reflections ("prel" in the diagram below), and three times does he interrupt himself with disclaimers ("d") intended to justify the oddity of a biographic undertaking in view of his world's avowed disinterest in the individual. The disclaimers grant readers a glimpse into the ambiguous entity implied in the ponderous pronoun "we": whereas "our intention" is certainly none other than the (presumably singular) chronicler's own, "our intellectual life" turns from him to the institution he represents. Without even once mentioning the name Castalia, the chronicler reminds us already on the first page of the tenets and self-definition of this "Order" or "hierarchy," to which he refers as to a well-known institution ("C").

The actual introduction to the glass bead game ("gbg"), so effusively announced in the heading but repeatedly delayed, does not get under way until another lengthy disclaimer has been delivered. In its three components, this time united in one block ("ddd"), the chronicler denies any intention to provide a satisfactory clarification, complete history, or textbook exemplification of the game. Now that he has established what his text is *not*, he begins his explanation. He outlines the game's state in his time, explains its origins in the intellectual attitudes of preceding historical eras, and tells the legend of its fictional inception.

Each of these segments refers explicitly to music as the purest expression of human emotions and to mathematics as the purest of the abstract sciences, stating that the meeting of the two ("m+m") caused the germinating spark for the game. The second segment includes a long quotation from Nicholas of Cusa's quasi-Socratic dialogue *The Layman on Mind* ("q1"),[21] which ends with the idea of universal harmony, realized when the Mind perceives the phenomena of the world through numbers and geometric figures. In the third segment, a meditation on the mythical powers of music establishes the introduction's main motif ("mus"). The section closes with musings about

[21]This quotation, from § 125 of the Latin *Idiota de mente*, can be found in a slightly different translation in Jasper Hopkins, *Nicholas of Cusa on Wisdom and Knowledge* (Minneapolis: The Arthur J. Banning Press, 1996), "The Layman on Mind," chapter 9, p. 569.

the power of music as revered by the ancients ("a") of Greece and China. China with its culture and philosophy is briefly mentioned three times in this section; it will become the introduction's second motif ("Ch").

FIGURE 7: Thematic analysis of the five sections in "A General Introduction ..."

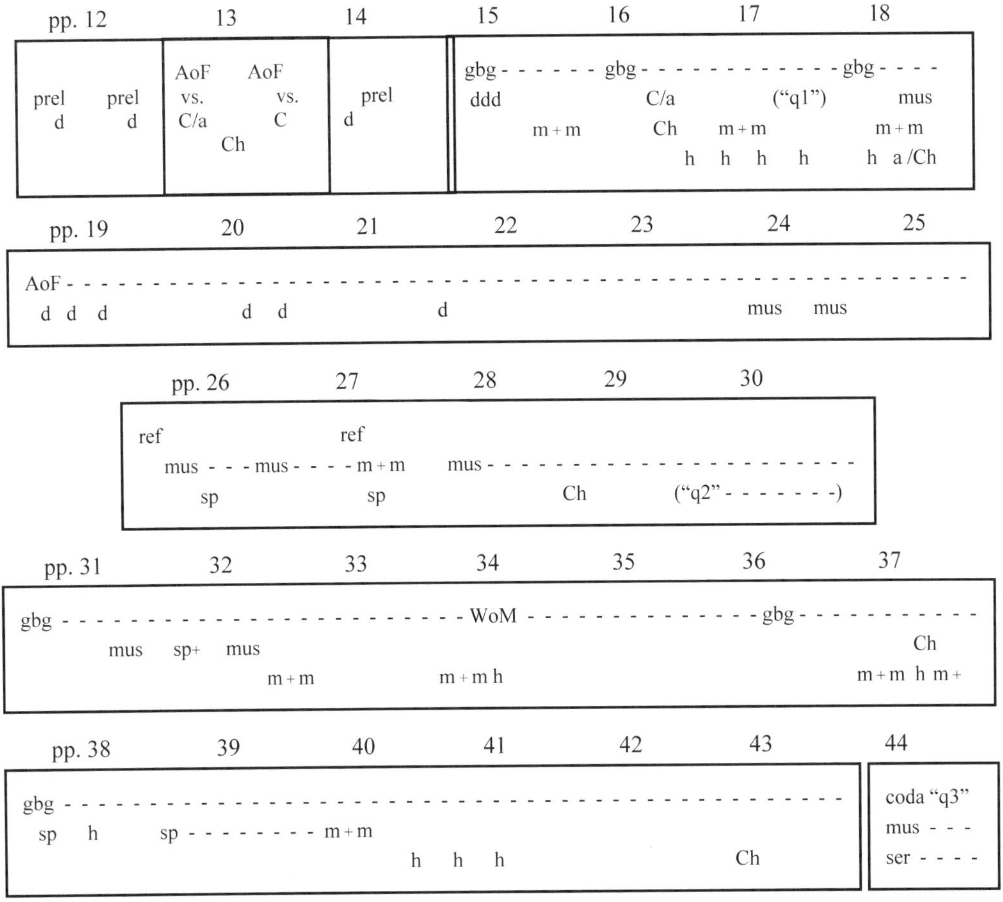

thematic material

- AoF = theme 1: Age of the Feuilleton
- gbg = theme 2: glass bead game
- ref = transitional topic: intellectual reformation
- WoM = supporting theme 2: the World of the Mind
- mus = motif 1: music
- Ch = motif 2: Chinese wisdom
- sp = motif 3 spirituality (the Journeyers to the East)
- /ser = serenity

supporting material

- prel = preliminary remark
- d = disclaimer
- C = Castalia
- a = ancient high cultures
- m + m = music + mathematics
- q = quotation
- h = universal harmony

The only portion of this section of the chronicler's introduction that is cast in bipartite form is the initial slur at the era of decay out of whose morass the new order grew: the Age of the Feuilleton ("AoF"). Twice and from two only slightly differing angles, this era is presented as unfortunately interpolated between the *ancient* high cultures in East and West (including their worthy heirs in subsequent centuries) and the *modern* high culture, i.e., the still unnamed Castalia.

The second section, ending on p. 25 with "a new organization," develops the first of the two themes the chronicler had introduced earlier: the "Age of the Feuilleton" and its superficiality, so incomprehensible to the ascetically erudite Castalians. The era's characteristics are described with a mixture of relish and horror. The contempt is in no way mitigated by the two threesomes of disclaimers with their phony warning against the unfairness of judgments made with the benefit of hindsight. The attempts at an historical account are allegedly based on the research of a literary historian named Plinius Ziegenhalss—a man whose vernacular last name, meaning goat's neck but hissing from its uncommon spelling with a double-s, suggests a pedigree of rural Alpine stock, whereas his Latinized first name suggests his parents' ambition to rear a classicist. The diatribe is in five segments, each full of hypocritically veiled scorn. It begins with the moderns' impression of the feuilletonistic age, backs it up with a narrative of the gradual decline of intellectual life during the centuries that preceded the era, and then offers a long description of some of its most astounding products. This description is interrupted for a lament over the tragedy supposedly felt by some of the citizens—a tragedy that could only be endured with the help of childish games such as crossword puzzles. The treatise ends with the chronicler's generous acknowledgment that the late feuilletonists' frequent despair and cynicism must be interpreted as first glimpses of a new awareness.

The central section of the "general introduction," ending on page 30 with "... let us return to our subject," bridges the two themes by reporting on the decades of transition following the Age of the Feuilleton and the eventual reformation of the life of the mind ("ref"). This reformation happened under the intellectual guidance of musicologists and the spiritual influence of the Journeyers to the East. Spirituality ("sp"), the composition's third motif, appears here primarily in its relationship to music, as the chronicler launches a long treatise on the Journeyers' musical tastes. He extols the virtues of classical Western music (particularly that of the 16th, 17th, and 18th centuries), recalls the symbolic role music played in ancient China with regard to the ethical state of the country and its ruler, and reflects on the connection between music and magic. As in section 1, the second of the three topics is substantiated with a long quotation, the above-mentioned passage from

The Annals of Lü Buwei ("q2"), centering on the concept of music as founded on the harmony between heaven and earth. Chinese wisdom is herewith firmly established as the second motif.

Section IV, ending on page 37 with "the history of our beloved Game," is wholly devoted to the secondary theme, which is explored in its two complementary aspects, the general and the specific: the new World of the Mind ("WoM") that resulted from the intellectual reformation, and the glass bead game as its stellar manifestation. The explanations of the game are presented in three segments, covering the evolution of its tools, its subsequent refinement, and its eventual universalization. All three of the movement's motifs are present, with a clear prevalence of motif 1, music.

The final section wraps up the discussion of the glass bead game. The chronicler reports on the game's evolution into a cult, its spiritualization and ritualization, its role in the epoch-transcending search for universal harmony and the synthesis of humankind's knowledge and wisdom, its relation to faith in general and to the Catholic Church in particular, and its ultimate institutionalization. Music as the main generating power at every stage is given full credit, as are the spirituality originally infused by the Journeyers to the East and the wisdom developed within the Chinese tradition.

In the "coda," the chronicler abandons his goal to offer an encompassing account of all aspects of the glass bead game. Summarizing it succinctly as "primarily a form of music-making," he links the victorious secondary theme to the composition's dominant motif. After another extensive quotation—Knecht's hymn to the spirit and piety of classical music, allegedly taken from a transcript made by a student—the chronicler concludes with an abrupt remark that leads back to his preliminary reflections. His manner here is so blatantly awkward that one must suspect him to be taking pains to distance himself from his hero's all-too-glowing homage to music, which stresses not only music's lofty links to morality and order, but also its power to express destiny, courage, defiance, and "superhuman laughter."

Beyond themes, motifs, formulas, and cadenzas, the thematic material in the introduction contains a fifth component: a hidden thread of terms referring to the idea of universal harmony ("h"). Its manifestations range from the name Pythagoras to terms like "oneness," "analogy," "correspondence," "*unio mystica*," "synthesis," "alchemy," etc. The two climactic quotations from Nicholas of Cusa and the Lü Buwei both contain references to the underlying thought, whereas Joseph Knecht's words allude to it only in metaphorical form as they prefigure the tension he will have to address.

In view of its five-section design, its framing, and the "cadenzas" of its three long quotations, one might describe the chronicler's introduction with words similar to those the narrator will use much later to characterize a glass bead game composition. Speaking of the entry Joseph Knecht submits to the annual glass bead game competition of the Waldzell elite shortly before his election as *magister ludi*, he calls it a game "of transparently clear, classical composition and strictly symmetrical development in the vein of the old masters" (p. 196). At the same time and with respect to its thematic development, the chronicler's "game movement" is curiously symptomatic of the story it prefaces. A synthesis of the two contrasting themes established in section I is not only never achieved—it is not even attempted.

The chronicler seems to perceive Castalia's glass bead game, along with the World of the Mind of which it is the esoteric symbol, as a lotus rising from the mud of a doomed and therefore overcome culture. Beginning with section III, he disregards the dialectical goal typical for bi-thematic structures —the striving toward synthesis. Instead, his treatment of the material is reminiscent of instrumental works in the Italian *canzona* tradition of the 16th and 17th centuries. These pieces also display clear sectionalization, and their themes often grow out of one another, with the result that there is neither a recapitulation of the original juxtaposition nor a concluding synthesis. It would seem that the chronicler had modeled the "musical structure" of his introduction after the keyboard and lute pieces that play such an important role in the lives of his tale's characters. Knecht's friend Carlo Ferromonte authors a *History of Styles in Sixteenth-Century Lute Music*; Knecht spends his first semesters in Waldzell studying the keyboard music of Purcell and his contemporaries; the Music Master treats Knecht's nervous exhaustion during the debating match with Designori by involving him in a discussion about the origin of sonata form; later, Knecht researches sonata form in Froberger; Father Jacobus seeks refreshment every evening by playing 17th-century keyboard sonatas on his piano; when Knecht seeks to alleviate the by now older Designori's psychological stress, he performs for him the Purcell sonata that had been Father Jacobus's favorite; and for his future pupil Tito, he analyzes a Scarlatti *Andante*.

It is tempting to try a symbolic reading of the musical form adopted in this prose section. A *canzona* or early sonata movement seems to begin with two themes. But rather than giving them equal weight, the composer soon allows the secondary theme to outshine the first and render it obsolete. In this light, the "musical" structure of the *Glass Bead Game* introduction functions as a prefiguration of what will be revealed as Castalia's great danger in the era of the *magister ludi* Joseph Knecht: the self-congratulatory aloofness of the World of the Mind vis-à-vis the world-at-large.

Part II
ARITHMETIC

Plate 2: *Typus arithmeticae*: arithmetic personified as a woman
Woodcut from Gregor Riesch, *Margarita philosophica* (Freiburg: Schott, 1503)[1]

[1]Robert Lawlor, who reproduces the woodcut on p. 7 of his *Sacred Geometry: Philosophy and Practice* (New York: Crossroad, 1982), includes in the comments he appends the following observation: "On her thighs (symbolizing the generative function) are two [numerical] progressions. The first series, 1, 2, 4, 8, goes down the left thigh, associating the even numbers with the feminine, passive side of the body. The second series, 1, 3, 9, 27, goes down the right thigh, associating the odd numbers with the masculine, active side, an association which goes back to the Pythagoreans, who called the odd numbers male and the even female. The Greeks called these two series the *Lambda* [...] On the woman's left sits Pythagoras using an abacus system for computation. [...] Boethius sits on her right using Arabic numerals [...].

The Eternal Realm of Numerical Relations

Whether Pythagoras really sat experimenting at the monochord when he found the numerical ratios of consonant intervals or whether he derived them intellectually, he is credited with determining the four principal intervals in music as well as founding the belief that the harmony of the universe is inherent in the mystical power of numbers. His names for the basic intervals, built with the Greek prefix *dia* for "between, through, across," have long functioned as a secret code for the entire concept. They are *diapason* for the octave, produced by strings with lengths in the ratio 1 : 2, *diapente* for the fifth (2 : 3), and *diatessaron* for the fourth (3 : 4).

These intervals can be built using strings with relative lengths of 1, 2, 3, and 4. Consequently and independently, the numerical sequence 1 2 3 4 was reputed to embody a spiritual value for Pythagoras and his followers. Speculation was nourished particularly by a specific figure of ten dots, arranged in an equilateral triangle with the dots seeming to form the equation $1 + 2 + 3 + 4 = 10$. In a spiritual reading of this equation, "1" refers to the essence of Divinity; it expresses itself through 2, duality, the first manifestation of spirit-and-matter; together, they produce 3, the animating "soul" of the cosmos; 4 represents the unfolded cosmos in its physical aspect—what humans perceive with their senses; and 10 refers to the whole as a functioning organism. This figure, called the *tetraktys* (the Greek word means something like "fourness"), can be taken to symbolize the first appearance of the cosmic monad of consciousness and the emanation from it of successively materializing aspects, forming the universe as we know it.

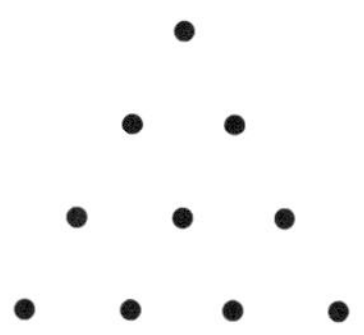

The Pythagoreans believed that numbers are self-existent, eternal, and immutable—hence, divine. All objects in the universe depend upon the relations between numbers. Pythagorean initiates were required to say a secret prayer or oath to the *tetraktys* when they began their three years of silence as novices: "Bless us, divine number, thou who generatest gods and men! O holy, holy *tetraktys*, thou that containest the root and source of eternally-flowing creation!"

Another reading of the diagram that has been suggested draws on initiate knowledge. From the first centuries C.E., scholars of Pythagoreanism have argued that the simple translation as 1 : 2 : 3 : 4 is probably too direct considering the sacred value attributed to the *tetraktys*. If Pythagoreans swore by the "holy tetraktys" as other brotherhoods swear by a god, they must have seen much more in it than an easily countable number of dots read as an equation, even if the resulting numbers symbolized a metaphysical belief.

A first cue to a possible secondary, veiled meaning comes from a work of art created almost 2000 years after Pythagoras's lifetime. The famous fresco *The School of Athens*, with which Raphael (1483-1520) decorated one wall of the Vatican's *Stanza della Segnatura,* shows Plato and Aristotle in the center and Pythagoras in the left foreground. The latter looks at a tablet presented to him by a young person, which shows a diagrammatic representation of a lyra juxtaposed with the *tetraktys*. As mentioned earlier, the lyre's four unchanging strings were tuned to the ratio 6 : 8 : 9 : 12. Raphael gives these numbers in the top row. Greek words written into the curves (2 x *diatessaron*, 2 x *diapente*, 1 x *diapason*, here replaced by the English terms) name the musical intervals at which these strings will sound. Below the schematic lyre appears a *tetraktys*, made up not of dots but of vertical slashes. A separate segment holds an X, presumably for the special number 10.

FIGURE 8: Schematic representation of a detail from Raphael's *School of Athens*

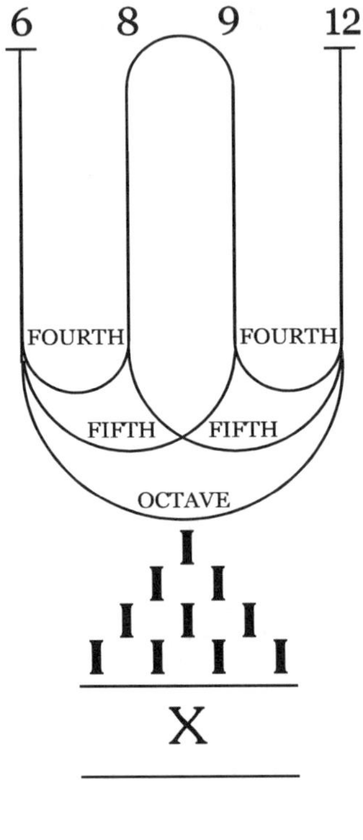

Scholars of Greek thought have read this depiction as a confirmation that the ostensible numerals 1 2 3 4 simultaneously served as a veiling screen for the secret sequence 6 8 9 12. This is in fact the reading that Nicomachos of Gerasa gave in the 2nd century C.E., albeit without any explanation. Kepler, too, discussing the *tetraktys* extensively in Book III of *Harmonices mundi*, refers to the tetraktys in its translation as 6 : 8 : 9 : 12.

Another quasi-esoteric application of numbers, also introduced in Nicomachos's influential book on arithmetic, relates to the "means" that mediate between two numbers. As the ancient Greek thinkers defined them and philosophers still hold, the proportions that determine these means are threefold: there is the arithmetic mean, the geometric mean, and the harmonic mean. The first and the last are manifested in both interpretations of the *tetraktys*. For the arithmetic mean, which simply halves the distance between a pair of numbers, see 1–**2**–3 / 2–**3**–4, but also 6–**9**–12. For the harmonic mean, in which the first amount is in proportion to the second as the second is to the third, see 6–**8**–12, where 8 exceeds 6 by one third of 6 (= 2) while it lacks one third of 12 (= 4); the same holds for 1–**2**–4.

Ostensibly the result of mere arithmetic operations, the means bring us back to the other sacred diagram of the Pythagoreans, the *lambdoma* matrix.

- The arithmetic mean (x = a + c divided by 2) manifests in all left-pointing diagonals in equally distanced values; see 1–**2**–3, 1–**3**–5, 1–**4**–7, 2–**3**–4, 2–**4**–6, 2–**5**–8, 3–**5**–7, and so on in every parallel.
- The harmonic mean (y = 2ac divided by a + c) is found in all right-pointing limbs in any group of three successive values; $^1/_1$–$\mathbf{^1/_2}$–$^1/_3$, $^1/_2$–$\mathbf{^1/_3}$–$^1/_4$, $^1/_3$–$\mathbf{^1/_4}$–$^1/_5$, and so on without exception.
- The geometric mean (z = square root of a x c) occurs in all pairs that lie at equal distance on either side of the vertical; the mean is here always an instance of 1.

This, then, is the reason why Boethius referred to the *lambdoma* matrix, developed in the musical context, as an "abacus." In secret teaching, it was regarded as "a representation of all the measures of the created world."

FIGURE 9: The common means in the *lambdoma* matrix

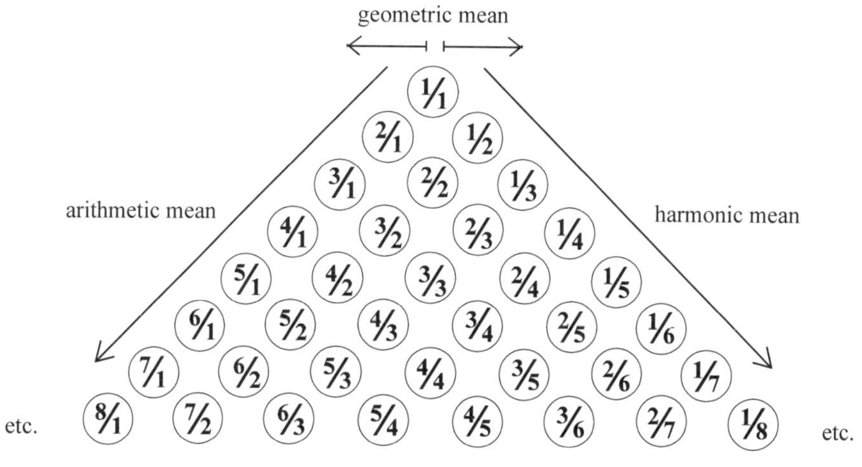

Another "mean" was discovered much later. In 1202, the Italian mathematician Leonardo Pisano (ca. 1180-1250) wrote a systematic introduction to arithmetic and Indian-Arabic numbers, *Liber abbaci*, the book of calculating tools. In Volume II he introduced the first numerical series ever discussed theoretically, which after his nickname, Fibonacci, became known as the "Fibonacci series." The series is composed in such a way that, beginning with the pair 0–1, each new number results from the addition of the last two (0, 1, 1, 2, 3, 5, 8, 13, 21, 34, 55, 89, 144 ...). The ratios of each pair of successive numbers circle an irrational value known as the Golden Section—Kepler's *proportio divina*—on which they converge alternately from below and above, coming closer as the series unfolds:

FIGURE 10: The Golden Section (also known as *phi*), 0.6180339887...

1:2	2:3	3:5	5:8	8:13	13:21	21:34	34:55	55:89
0.5	0.666...	0.6	0.625...	0.615...	0.619...	0.617...	0.61818...	0.61797...

This irrational value must be considered as the fourth member in the group of "means" discussed above: it bisects a geometric length (or temporal duration) in such a way that the ratio of the whole to its longer segment is the same as the ratio of the longer to the shorter segment.

The *lambdoma* matrix, the arithmetic, geometric, and harmonic means, the *tetraktys* with its two readings, the Fibonacci series, and the Golden Section were considered to constitute the entire array of mental and spiritual tools at the basis of divinely created nature and humanly created art.

Finally, the force of numbers also distinguished the four sciences of ancient Greece thinking, comprising what the early Middle Ages called the *quadrivium* (arithmetic, geometry, music, and astronomy), as distinguished from the remaining three components of the seven liberal arts, the *trivium* (grammar, rhetoric, and logic). The Pythagoreans perceived the four sciences as essentially governed by numbers. When they divided them into two pairs, they did so not with reference to subject matter but with regard to the way numbers applied:

- Numbers defining *amount* determine arithmetic and music, whereby arithmetic works primarily with separate amounts, while music is typically concerned with relations between values.
- Numbers defining *size* rule geometry and astronomy, whereby geometry works with abstract, static shapes and bodies, while astronomy observes shapes and bodies that are concrete and in motion.

The main dichotomy informing these distinctions is that between ideal objects and objects accessible to the senses, i.e., between *noeta* and *aistheta*. Both play a role for Kepler as they do also for Hesse and Hindemith.

Dramatic Analogy and Musical Proportion in Hindemith's Opera

In his libretto, Hindemith divides the plot of *Die Harmonie der Welt* into five acts but not into scenes; in the score, he marks changes of musical material with rehearsal cues but not with headings. On a prefacing page in the score, he lists 14 different locales, each followed by a dotted line as if he had planned to give page numbers which he then forgot. This table of venues is usually interpreted as a list of "scenes" and the dramatic action accordingly subdivided as 4 + 4 + 2 + 1 + 3—a somewhat astonishing assumption given Hindemith's well-known proclivity for order and symmetry.

A division into dramatic segments is no doubt useful, but it is not clear that scenery changes are the best criteria. I propose an alternative grouping based on content, which highlights fascinating correspondences in the structural layout. As shown in the table below, where bracketed numbers refer to Hindemith's musical cues, acts I and II are built in striking analogy. The composer plays with corresponding musical forms and dramatic content: He matches analogs: a song with a rondo and propaganda with status-building, but also opposites: grief with hope and vision with illusion. Emphasis on the protagonist's social relations in act I is juxtaposed with focus on his intimate concerns in act II.

I	II
[1] Tansur with crowd + Ulrich *propaganda* (SONG)	[9] Tansur with paupers + Wallenstein *status-building* (RONDO)
[2] Prague: Wallenstein +Tansur	[10] Linz: Ulrich + students
[3] Mother Kepler's grave-digging *reproach for intellectual conceit* STROPHIC DUET with Christoph Kepler	[11] Pastor Hizler's outrage and ban *reproach for spiritual conceit* STROPHIC DUET with Johannes Kepler, + ensemble/choir passages)
[4][5][6] emperor–Kepler, + mother Kepler's *vision*	[12][13] baron–Kepler, + Ulrich's *illusion*
[7] daughter–Kepler: *affection* + *grief* over lost loved ones	[14][15] Susanna–Kepler: *affection* + *hope* for a shared future
[8] Kepler with Ulrich REPRISE: 2 stanzas of INITIAL SONG, sounding through the window	[16] Tansur recruiting soldiers REPRISE: episode of INITIAL RONDO + reminiscences Kepler/Susanna
Kepler, professional and family man	*Kepler, believer and soul mate*

Similar analogies pertain in acts III and IV: parallel musical structures are complemented with corresponding dramatic patterns. After the opening rondos, the protagonist's mother and last employer respectively—two persons who dominated Kepler's life at its opposite ends and who both craved glory and wealth through self-assertion—seek to misuse his theses about universal harmony for their own ends. Once again Kepler is portrayed from two contrasting angles: capable of drawing boundaries in act III, resigned in act IV.

III	IV
[17][18] "moon RONDO" Kepler + mother, wife, daughter	[22] "palace RONDO" Wallenstein + guests, Ulrich, Kepler
[19][20][21] "trial rondo" a *superstitious* mother Kepler *solicits Kepler's collaboration* in her approach to "world harmony"	[23][24] conversation a *superstitious* general Wallenstein *solicits Kepler's collaboration* in his approach to "world harmony"
Kepler rejects, embarrassed	*Kepler acquiesces, embarrassed*

The third juxtaposition, which concerns only dramatic analogies but no musical correspondences, occurs *within* act V:

[25] Sagan: Susanna, Ulrich re: Kepler's avoidance of home; *Kepler's loss of personal roots*	[26] Regensburg: emperor, electors re: Wallenstein's dismissal; *Kepler's loss of professional basis*
[27] Kepler's *nightmarish damnation* + *descent* into death	[28] Kepler's *dreamlike transfiguration* + *ascent* into cosmic allegory
Remorse over strife and delusion	*Joy over serenity and higher insight*

The carefully designed structure of analogs seems altogether appropriate for a work whose protagonist strives to ascertain harmonic proportions in all spiritual and material aspects of Creation. Guided by this table, the dramatic layout can be read as a threefold juxtaposition. This should be kept in mind when approaching Kepler's poem of mourning, which plays such a crucial role for the spiritual dimension of the opera (see part VII below). There, the first $\frac{2}{3}$ of stanzas 1 and 2 each present an external/internal contrast similar to that shown for acts I : II and III : IV, followed by a different dualism captured in the final $\frac{1}{3}$ of *both* stanzas, which corresponds to that *within* act V.

Encouraged by these discoveries, one wonders whether the operatic portrayal might not aim for ratios as simple and perfect as those Kepler sought to ascertain in cosmos and nature. As Hindemith lovers know, this composer can be trusted not to eschew such a challenge. And indeed, a play with proportions can be detected. Whether intended by the composer or sought by the eager analyst's eye, this play is intriguing, its ratios as intricate as they are simple. (The equations I give below are based neither on counted pages or measures, thus not on visual proportion, nor on the playing time as realized in the recording,[1] but on the ideal duration of each musical segment according to Hindemith's metronome markings and changes in tempo and meter.[2]

To begin, the acts display harmonious proportions:

act I : act II	=	3 : 4	(fourth)
act I : act III	=	4 : 3	(fourth)
act I: act IV	=	2 : 1	(octave)
act I : act V	=	8 : 9	(whole tone)
act II : act III	=	5 : 3	(major sixth)
act II: act IV	=	8 : 3	(octave + fourth)
act II : act V	=	6 : 5	(minor third)
act III : act IV	=	5 : 3	(major sixth)
act V (earth : cosmos)	=	5 : 3	(major sixth)

The sum of all dramatic segments addressing an earthly search for harmony relates to those expressing its heavenly realization in a perfect ratio:

act I – act V middle : overture + cosmic finale = 8 : 1 (three octaves)

Similarly harmonious proportions prevail on the next level. For instance, in each of the two rondos of Act III, a refrain alternates with "common" and "unusual" episodes. In the moon rondo, the refrain itself is bipartite, comprising a children's song and a lunar reply; the unusual episode is a hymn. The ratio of these components, again with regard to duration, is as follows:

stanzas of lunar replies	: stanzas of children's song	=	4 : 5
sum of all regular episodes	: stanzas of children's song	=	3 : 4
hymn	: stanzas of children's song	=	3 : 4

In the "trial rondo," the sum of all refrains (which establish mother Kepler's attitude during the court proceedings) relates to the sum of all episodes as 5 : 8. The regular episodes (which contain the accusations brought forth by magistrate and witnesses) outdo the first unusual one—a collage segment that leads to mother Kepler's acquittal—at a ratio of 9 : 8. Most intriguing is the observation that the sum of all public accusations against mother Kepler (including her subsequent acquittal) is balanced 1 : 1 by her real madness: her railing against her ungrateful children and her subsequent "revelation rondo," in which she unfolds her plan and shows that her belief in her supernatural powers far outstrips any official suspicions.

[1]I am referring to the 2002 CD recording conducted by Marek Janowski (Wergo 6652 2).

[2]No absolute certainty for Hindemith's intended durations is available even for diligent, calculator-equipped scholars, owing to the indefiniteness of *ritardando*, *accelerando*, and *ad libitum* passages. But, to risk an ambitious comparison: my values come as close to the stated proportions as do Kepler's inductively derived angular speeds of the planets, laid out in Book V of his *Harmonices mundi.* (More on this in part IV below.)

Tetraktys, Pentagram, and the Disharmonious SEVEN

Like many composers, Hindemith often thinks in countable units. These may be the stanzas in a song, the subject entries in a fugue, the insertions into an otherwise straightforward development, the pitches in a melodic unit or chord, etc. Befitting the subject matter, a considerable portion of the conspicuous play with numbers is based on the four numerals in the secret *tetraktys*, 6 8 9 12, the sequence the Pythagoreans deemed perfect because it embodies the arithmetic and the harmonic mean as well as the four basic intervals.

Sets of TWELVE occur on SIX prominent occasions in this opera. In many cultures and thought systems, the number 12 symbolizes completion; an odd thirteenth element is perceived as outside this order and thus disharmonious, but may also serve to draw attention to the order. Hindemith employs the number 12 with emphasis on its ideal divisions (6 + 6, 8 + 4, and 9 + 3); he also explores subtly asymmetric subdivisions for double the amount, 24.

As already mentioned, the overture introduces the subject matter: the quest for universal harmony. The pervading motif is heard 12 times in unison on the tonic E; an odd additional entry interrupts the serenity after the 9th regular statement with a somewhat dissonant four-part parallel. The fugato in act II, in which Linz's Lutheran congregation queries whether the pastor has the right to judge and ban a pious man, consists of 12 entries (6 of them vocal, the other 6 instrumental); in the codetta, almost concealed in a longer vocal line, Kepler adds a 13th entry when he asks the pastor: "Do you realize the consequence of your step?"—the consequence being disharmony. In the moon rondo of act III, Kepler's hymn to the divinely created cosmos is a passacaglia-fugue with 12 entries, 6 in the first half with Kepler's monologue, the other 6 (in the same key sequence) in the structurally analogous second half. Later in the same act, the trial rondo—spiritually a diametric contrast to the hymn—turns on a refrain conceived as a different kind of unusual fugue (more on these later). This one features 24 subject entries altogether, grouped as 7 + 5 + 3 + 1 + 2 + 2 + 4 (or [12] + [4]+[4]+4).

At the end of act IV, Wallenstein seeks Kepler's collaboration for his intended political application of "world harmony." A segment filled with shameless flattery—acknowledged by a wary Kepler with the words "Thus spake the tempter"—is cast as a fugato with 12 subject entries. This time, 8 instrumental entries balance 4 vocal ones. In one of these, Wallenstein asks grandiosely: "Can greatness thrive without sacrifice?"; in the three others, Kepler's above-quoted line about the tempter is musically linked to his incredulous "And I am the one on whom you are counting as a helper?" at the outset and his unhappy "The sacrifice: too great it already seems today!" toward the end.

Ironically, the most perfect realization of the number 12 occurs when a plan to make all things work out "harmoniously" is taken to its irrational extreme. Mother Kepler's "revelation theme" in the trial rondo, in which she lays out how she will save herself, Kepler, and the rest of the world, uses all 12 semitones in a phrase spanning 12 half-note beats, as if to underline musically that hers is a truly complete system of (bizarre) thoughts.

In addition to their occurrence in the context of an emphatic TWELVE and as symbols of specific dramatic situations, the numbers of the *tetraktys* also shape the overall occurrence of elements. Each of them is expressed in one of the music's most notable components: there are TWELVE stanzaic forms,[3] NINE instrumental interludes,[4] EIGHT passages strung together with a repeated-chord accompaniment,[5] and SIX cases of intramusical quotation.[6]

The number FIVE has a special place in Pythagorean thinking. In plane geometry, the pentagram is connected to the only irrational proportion, that of the Golden Section; in spatial geometry, the only one among the five Platonic solids whose faces are pentagons—the dodecahedron—is also set apart by not being philosophically linked to one of the four elements of earthly reality (fire, air, water, and earth) but to the starry sky.

On the two occasions in *Die Harmonie der Welt* when Hindemith shapes musical material with the help of the figure 5, he is evidently thinking of just this irrational aspect. In act I, mother Kepler sneaks into the cemetery to excavate her dead father's skull, which she believes she needs to cure her son of the "intellectual conceit" of his scientific approach and open his mind to

[3](1) Tansur's propaganda song in [1]/[8]; (2) mother Kepler's grave-digging song in [3]; (3 + 4) Emperor Rudolf's ariettas in [4 B] and in [6]; (5) the song of mourning sung by Kepler and his daughter in [7 B-E], (6) Ulrich's arioso in [8 A-B] / [8 E]; (7) baron Starhemberg's arioso in [12]; (8) the arioso in varied stanzas expounding Kepler's creed in [15]; (9) the song shared by Susanna and Kepler in [14], (10) the song Kepler's young daughter sings to the "little man in the moon" in [17]; (11) the electors' war song in [26]; (12) the varied strophes in which Pastor Hizler couches his accusation against Kepler in [27].

[4](1) at [2 E-G]; (2) at [5 L]; (3) at [6 C]; (4) at [9 S-W]; (5) at [13 F-G]; (6) at [19-19 C]; (7) at [22]; (8) at [26]; (9) at [28].

[5] (1) at [4 -4 A]: 11 measures over a 6-pitch chord; (2) at [8 D]: 9 measures over a 4-pitch chord + flute figure; (3) at [11 C-I]: 58 measures over a unison pitch; (4) at [12]: 8 measures over a homophonic woodwind figure; (5) at [14]: 13 measures over a homophonic woodwind figure; (6) at [20]: 42 measures over an 8-part tutti collage; (7) at [23]: 50 measures over a minor-seventh interval; (8) at [24 M-P]: 44 measures over a minor-seventh-based figure in the timpani.

[6]The melodic materials linked to Kepler's thoughts about death, his spiritual creed, his praise of cosmic order, and the pastor's doctrinal righteousness are each quoted once, whereas the music for Susanna's declaration of her harmonic goals recurs twice. These internal quotations will be discussed in later chapters.

a larger understanding. As she begins digging, she sings a song in 5 stanzas, each consisting of 5 segments. In the subsequent scene, Emperor Rudolf muses about the realm of beautiful illusions in which he lives as a sovereign —a "phantom world," as he calls it—and expresses his desire "to slip dreamily away to the indefinable sphere of starbright, slumbering wisdom." He couches his longing for cosmic solace in 5 statements of an ascending motif (shown in Ex. 7 above), which is sustained by 5 realizations of a 5-note chord.

Related to the figure 5 is the juxtaposition of 2 and 3. It comes in three guises: the two numbers may complement one another as parts of a larger 5; they may compound one another as divisions of 6; or they may display aspects of one another. The first is most notable in Kepler's hymn to cosmic order, which alternates measures in $\frac{2}{2}$ and $\frac{3}{2}$ time. The second manifests in several of the ostensibly simple musical structures; I will single out the case of the opera's three pairs of rondos to argue my point, but similar observations could be made with regard to the opera's three two-stanza songs, the dual aspect in the three marches, and the two passages of existential danger accompanied each by a three-part polymetric ostinato.[7] Finally there is the tonal aspect whereby triads—entities of three pitches in harmonic relation to one another—may oscillate in constant ambiguity.

Die Harmonie der Welt features six rondos of different scope and complexity. They come in three pairs. The first pair comprises the March that in [9] launches act II and the "ballet polonaise" that in [22] opens act IV. They are related with regard to their musical simplicity as well as by external factors: the persons involved and the locales at which the corresponding actions take place. Both center on Wallenstein, who is assisted both times by an eager Tansur. The earlier scene plays at the site of Wallenstein's future palace, the later one inside this palace. In the open-air scene, a mob of homeless people resent losing their makeshift dwellings but thank Wallenstein for the refund he offers them; in the ballroom scene, an assembly of elegant citizens and officers praises and thanks the successful general while silently resenting his grandiosity. The general, addicted to wealth and power, in both cases takes in stride the injury of those whom his ambition marginalizes.

The "moon rondo" and "trial rondo" form the second pair; they are as complex as those in the first pair are straightforward. Their musical material and part historical, part fictional background will be discussed in part III.

[7]These rather striking passages occur, one during the "trial rondo" of act III when mother Kepler faces accusations for acts she hasn't committed (though she does plan others), the other during the temptation Kepler faces in act IV when Wallenstein invites him to forfeit his idealistic quest for a utilitarian application of his ideas.

Their proximity in the opera's central act, their human link in the form of mother Kepler, and their shared preoccupation with the dark side of human life have a structural analog: both harbor a substantial episode whose musical features threaten to defeat the surrounding dance form.

The last two rondos are analogous, structurally in that both take the place of episodes *within* larger forms, and spiritually in that both emphasize an unrealistic dream, which one may imagine turning round and round in the respective person's head. Inside the "trial rondo" of act III, at the point immediately following mother Kepler's acquittal, her "revelation rondo" divulges ideas and plans that alone might have sent her burning at the stake. In the "ballet polonaise" in rondo form at the beginning of act IV, the small-scale rondo at [22 D-H] articulates Wallenstein's fantasy of a harmony that would result from the horrors or war, of tolerance and liberal thought arising from forced peace, and of joy defying death and sorrow.

Finally for the correlation between 2 and 3, tonal ambiguity in triadic melodies occurs 3 times in *Die Harmonie der Welt*. In each case, the music features two competing fundamentals (F and F♯) and two alternative modes (major and minor), resulting in an oscillation between 3 cross-related triads (F♯ major, F♯ minor, and a never fully realized triad on the fundamental F).

EXAMPLE 9: Three characters losing their footing

Rudolf II: Ein Tu - - - bus, der, wie es heißt, [4 b]
(A tubus, which, it is said)

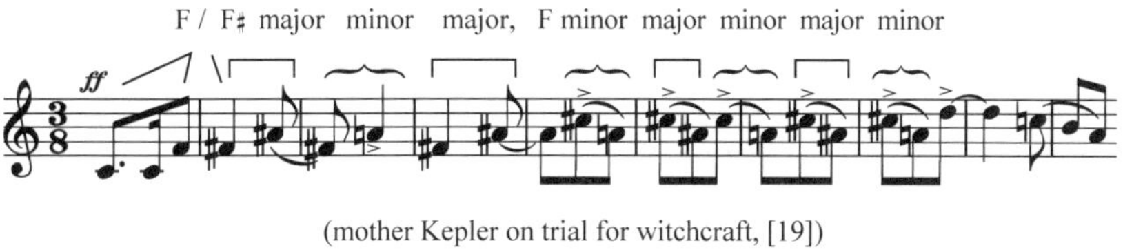

(mother Kepler on trial for witchcraft, [19])

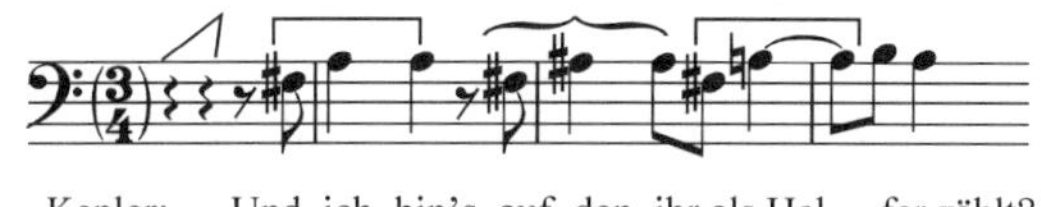

Kepler: Und ich bin's, auf den ihr als Hel - fer zählt? [24 F]
(And I am the one you are counting on as a helper?)

In all three cases, the tonal ambivalence occurs in a line sung by (or associated with) a person who is about to lose footing and succumb to forces that counteract wishful thinking and wise precautions. The emperor's first look through the new telescope will crush his cherished image of a simple and predictably ordered universe, and mother Kepler's resolve to disclose her secret plan will meet with her learned son's abhorrence and rejection. Kepler's acknowledgment that he is being recruited to do the maths for an ambitious military venture, along with his recognition that his quest for harmony is being coopted to provide vindication for the attack, will make him loathe—and lose—himself, as he believes to have forfeited his spiritual aim in the interest of providing for his earthly needs.

Another number that recurs conspicuously in various segments of the musical design is SEVEN. The connotations it evokes in both the biblical and the Pythagorean traditions are alarming. In the book of Revelation, 7 stands for "all" in the context of punishment: 7 seals unleash the entirety of man-made disasters, 7 angelic trumpet blasts call forth the divine retribution for human betrayal, and 7 bowls pour God's wrath over humankind, destroying a third of all that lives. In the Greek music-theoretical system of ratios expressed as partials and multiples of a fundamental tone, the number 7 is excluded as not fitting in with perfect harmony, deemed "out of tune" both then and now. 7 is also the only one-digit number by which a circle cannot be divided with purely geometric operations.

Hindemith employs the number 7 correspondingly, linking it exclusively to jarring situations. Tansur's propaganda song in act I, in which he combines scorn for gullible consumers of horoscopes with acid criticism of the monarchy, consists of 7 stanzas; Pastor Hizler's lengthy castigation of Kepler's alleged heresy in act II is cast in 7 phrases, and mother Kepler's "revelation rondo" in act III comprises 7 statements of the refrain phrase. The electors who in act V force the emperor to dismiss the war-and-power-hungry general Wallenstein sing their accusations to the tune of a war song from the 1620s which Hindemith set as a "theme with variations": the opening instrumental segment, played before the curtain is raised and setting the stage for the topic of war, together with the 6 stanzas of text in the variations result in another 7-stanza song.

The play with arithmetic proportions thus pervades all dimensions of the musical work, from the largest to the smallest: its dramatic layout, the temporal relationship of its segments to one another, the grouping and juxtaposition of compositional forms, the number of repetitive musical passages, the shape of meter, rhythm, and phrase length, and even the tonality.

Numerical Order in the Two Worlds of *The Glass Bead Game*

Hesse's compound work has two topics: the game that serves as the symbol of a World-of-the-Mind called Castalia, and a particular individual in that world, the outstanding *magister ludi* Joseph Knecht. The game, true to its nature as the realization of an ideal, is seemingly timeless and eternal—or so its adepts wish to believe. As outsiders we are aware that Castalia has not always existed; we also guess that it will most likely one day cease to exist. Yet based on data from within the story, all we know is that its inception must be imagined some time in the early 21st century and that by the time the chronicler put pen to paper, around 2400, Castalia is still alive and well. Knecht is placed into a particular historic era, but it is only vaguely specified as "around 2200." Knecht's life is not anchored in known years for birth and death—in fact both his childhood and his last days are shrouded in mystery, as the chronicler admits: "No knowledge has come down to us of Joseph Knecht's origins" (p. 47) and "of the Magister's last days, [...] we know no more [...] than every Waldzell student and could not tell the story any better than the *Legend of the Magister Ludi*" (p. 369).

On the surface, the story thus presents Castalia as a seemingly ageless institution, while its *magister ludi* undergoes a development. Beneath this surface, however, all is not that simple: there is by no means only one Castalia but three, and Joseph Knecht's maturation process is traced not in a single linear development, the consecutive events of a biography, but in three dimensions. This dual play of threefold manifestations can usefully be considered in parallel to Boethius's three manifestations of music:

- In correspondence with *musica mundana* (music as an all-pervading force in the cosmos) there is the idealized Castalia—envisaged in the motto and portrayed in the chronicler's account as unerringly maturing toward its lofty goal. In this Castalia, the various disciplines, then also mind and soul, and finally even the "World of the Mind" and the world-at-large are imagined as approaching perfect harmony. And there is Joseph Knecht's education in the elite schools, climaxing in his self-realization as a glass bead game player: an idealized life devoted to "the dream of capturing the universe of the intellect in concentric systems" (p. 16).
- On the level of *musica humana* (music as a principle unifying the human body and soul), there is Castalia as a reality that by the time the story is told has been in existence for several centuries, in which the chronicler and his contemporaries still believe, and whose recent reforms and improved usefulness to the outside world they credit to Magister Josephus's valiant act of conscientious rebellion. And there

is Joseph Knecht's maturation in a process of inner harmonization, achieved through the gradual integration of all the insights that five significant mentors and challengers prompt in him.

- On the level of *musica instrumentalis* (music as sung and played by humans), there is the all-but-ideal present in a Pedagogical Province that, in the 23rd century, has lost its essential connection with the world that spawned it and which it is meant to serve. And there is the sequence of Knecht's five consecutive lives in the course of which his consciousness, by way of mundane tasks, errors, suffering, and death, works toward a realization of its full potential.

Again on the surface, the narrative is presented as if consisting of THREE segments; the table of contents lists an introduction, the life of the *magister ludi*, and Knecht's posthumous writings as its major headings. At slightly closer inspection one discovers FIVE components, expressing different views of Castalia as was already mentioned. These are of strikingly different length and inner attitude and also represent different temporal perspectives.

- The motto, half a page long, is presented as excerpted from a treatise of the scholastic era; its ostensible purpose is to prove that centuries ago, far-sighted people already envisaged the eventual existence of an exercise based on glass beads and their spiritual value, believing it desirable though improbable. The stance is prophetic.
- The 33-page introduction, allegedly composed around 2400 C.E. and thus representing the opposite end of the temporal spectrum, posits the scholastic author's wish as a dream come true, explaining the spiritual-intellectual game's genesis and value in the form of a somewhat mystifying panegyric.
- By far the most substantial of the five components is the 380-page biography of the *magister ludi*. Personal detail is sparse and the narrative is forever interrupted for reflection—a credible enough outcome if we are to believe in a biographer allegedly working with archival material gathered at an institution that decries individual idiosyncracies or adventures and values only "the greatest possible integration into the generality, the greatest possible service to the superpersonal" (p. 12). In style and form, then, the biography is a piece of essayistic fiction.
- The poems Joseph Knecht wrote during his student years are the only component in which the (fictional) time of composition corresponds with that of the perceptions, thoughts, hopes, and fears expressed. An outlet for his nascent doubts about the Castalian order, they are cathartic and represent—emotionally though not grammatically—the first-person perspective.

- Finally, the three conjectural autobiographies, with 112 pages the work's second-largest component, originate from Knecht's years of free study and chronicle—in the world of the fiction—his repeated reflection on themes that define him, his growing self-knowledge. While their alleged author's time of writing covers at most ten years, the stories told span some 25 millennia.

All this can be detected after a cursory acquaintance with Hesse's work. Only repeated perusal and a focus on theme and structure will open readers' eyes to yet another plan. I wish to argue that Hesse's work consists of two interlocking strands, one addressing the idea and realization of a "World of the Mind" within the world-at-large, the other telling the story of a soul. Each of the strands in turn consists of five components. The one dealing with the "World of the Mind" discloses novel insights if read as laid out in analogy to a five-movement musical piece. Three of the "movements" are self-contained entities: the chronicler's introduction in part I of the book, Joseph Knecht's circular letter in chapter 11 of part II, and the Directorate's reply in the same chapter.[8] A forth component equals a musical piece whose initial section is set off from the rest by way of a change in tempo and character, as happens in some symphonic movements. Its opening section consists of two monologues (the young Knecht's and Designori's respectively, contained in the chapter "Waldzell"); the remaining four sections comprise a dialogue conducted some twenty-five years later by the same two men (constituting the major part of the chapter "A Conversation"). The final component in the five-movement composition is made up of Knecht's posthumous poems. Meanwhile, the other strand in the weaving, narrating the "story of a soul," consists of Joseph Knecht's five lives as prefigured in Hesse's sketch of 1931: his incarnations as rainmaker, father confessor, Indian prince, Swabian theologian, and Castalian.

While many Hesse scholars have commented upon the serial lives, the interwoven strand devoted to the impersonal theme of two worlds moving into increasing estrangement appears to have escaped their notice. This is surprising as the coexistence of the two interdependent threads informs Hesse's two-part title: *The Glass Bead Game // [Attempt at a description of the Life of the] Magister Ludi [Joseph Knecht].*

[8]It seems relevant in this context that chapter 11, containing the circular and its reply, was not written in sequence but much earlier. As documented in *Materialien* (vol. 1, pp. 43-44), Hesse sent chapters 1 and 11 to Suhrkamp in September 1938 for a possible preview publication in the *Neue Rundschau.* Volker Michels lists chapter 1 as having been written in January 1938, whereas in an entry for July 1937 he records the completion of "a new chapter of *The Glass Bead Game.*" It is thus possible that Hesse conceived circular and reply already in extension of the introduction and composed them before he started work on the Life proper.

Mind Against, With, or In Nature: a Dialectic Composition

As shown above, the chronicler's introduction not only exhibits typical features of a symphonic movement; it also satisfies the criteria of a classical glass bead game "in the vein of the old masters." I want to demonstrate that the same holds true for the larger entity: that the five-movement composition in strand 1 constitutes a "glass bead game" on the impersonal themes of Hesse's work, developing in a sequence of discursive steps that answer one another both within each "ludic" movement and in the whole.

Like the movement 1, movements 2, 3, and 4 of this "game" are also each designed in five segments, and all movements including the fifth are built on thematic material with five components: two contrasting themes and three motifs. The original main theme, the intellectual depravity of the feuilletonistic Age, and its antithesis, the World of the Mind with its glass bead game, recur as leading themes in movement 3; music (the introduction's motif 1) joins the glass bead game to form the dominant theme in movement 5, while Chinese wisdom and spirituality/serenity also recur in various guises.

Movement 2 turns on two themes that are implied but not verbalized in the first: the distance between the World of the Mind as it has evolved and the world as it continues to exist (th1: DIST), and the desired ideal of harmonization (th2: HAR). The themes are established and subsequently developed both separately and jointly; several attempts at a synthesis all lead to conflict (CONF). Among the recurring motifs, mutual respect (motif 1: MR) stands for a first step toward reconciliation, bitterness (m2: BITT) is the world's reaction to failed attempts at approximation, whereas serenity (m3: SER) characterizes the Castalians' effort to address and overcome the rift.

The movement's first section with its two monologues offers a kind of summary assessment, achieved in the two Waldzell debating partners' complementary contributions. In a letter to his mentor, the Music Master, Knecht tells about the misgivings Designori's criticism of his world cause him, revealing his nascent doubts whether the glass bead game is perhaps really no more than irresponsible though sophisticated play and Castalia, a sterile, parasitical, and conceited drone. Designori's farewell confession to Knecht was allegedly preserved in a transcript made by Knecht's friend Ferromonte. In it, the guest student from outside Castalia pleads for a recognition of the dangers inherent both in a world that overestimates intellect, sublimates feelings, and divorces itself from instinct, and in a world that advocates "naive living without discipline of the mind" and is "bound to become a mire into which men sink, reverting to bestiality" (p. 107). The two monologues are of exactly equal length (one page each; cf. pp. 96-97 and 107-108). They presuppose but do not dwell on the movement's two themes (the regretted distance and the desired ideal of harmony between Castalia and the world), and instead emphasize mutual respect.

FIGURE 11: Layout and thematic material in a rekindled debate

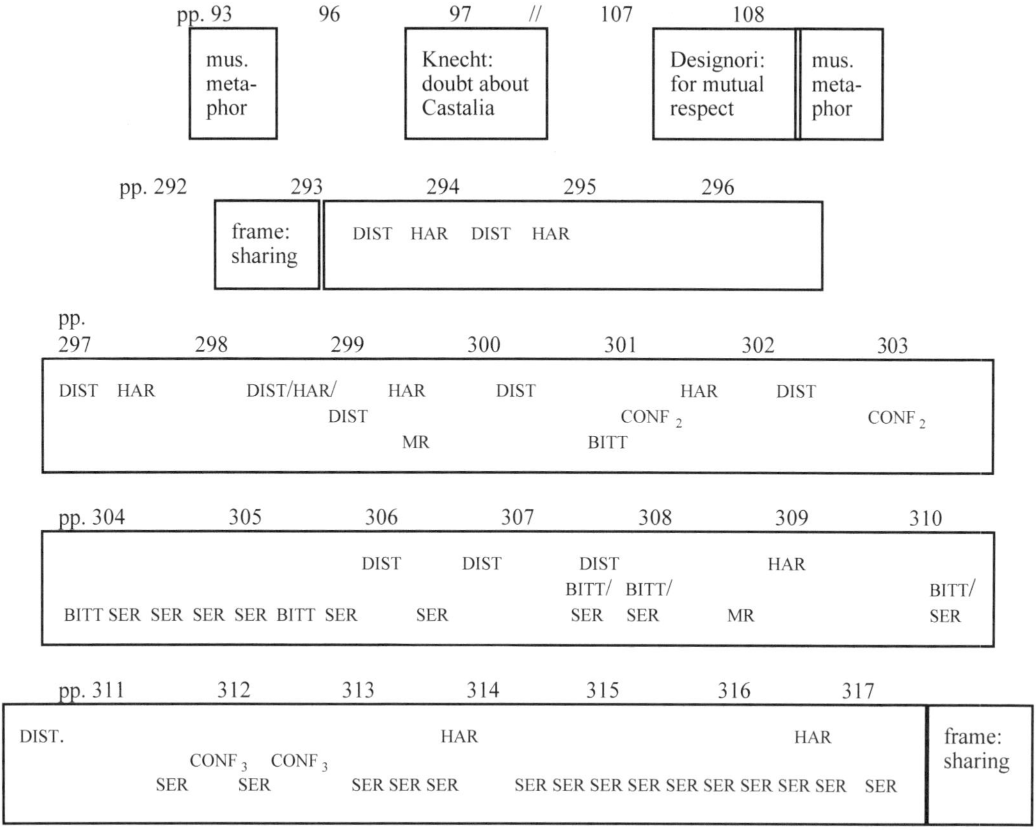

thematic material

DIST = theme 1: distance between the World of Mind and the world-at-large
HAR = theme 2: harmonization
CONF $_1$ = conflict at the attempt of bringing Castalian values into the world
CONF $_2$ = conflict at bringing the world's superficial interest into Castalia
CONF $_3$ = conflict reigning in the soul of individuals torn between both worlds
MR = motif 1: mutual respect
BITT = motif 2: bitterness
SER = motif 3: serenity

The sequel to this exchange, the dialogue occurring a quarter century later (in Hesse's chapter 9), comprises the movement's second to fifth sections. Section 2 begins with an exposition of the two contrasting themes. Each theme is then separately developed in a spirit of unmasking: Knecht initially attempts to make light of the perceived distance and idealized harmony and proposes a dialectic of comprehension. In its light, the belief that all human beings can basically communicate with one another, and its antithesis, that there are no two beings between whom complete understanding is possible, are both true. Designori moves the conversation from abstraction to experience. Debunking the fiction that synthesis is realized by guest students shuttling happily between elite school and worldly home, he tells of his failure to achieve harmony by bringing Castalian values and attitudes into the outside world (conflict 1). Section 3 opens with a restatement of the two themes. Once again each is then developed separately, but this time in close alternation and supported by motif 1 (mutual respect). The section concludes with another reminiscence of conflict, this time caused by the world's attempt at expressing its well-intentioned but superficial interest in Castalia's aims and attainments through brief visits and dabbling participation in its game. This second version of chafing contact (conflict 2) breeds bitterness, the movement's second motif.

Section 4 opens with a lengthy battle between the motifs of bitterness and serenity, which eventually leads to a restatement of the theme of distance. A new development, centering on the motif of mutual respect, closes with a reprise of the theme of harmony, now also played out before the backdrop of bitterness versus serenity. The final section 5 opens with the recapitulation of the main theme. A serene glance at the clash between actual distance and desired harmony reveals the third instance of the conflict, one that takes place on an individual's inner arena (conflict 3). The movement concludes with a recapitulation of the theme of harmony, surrounded by long bolstering passages in which the motif of serenity fills every crevice of the verbal texture.

The opening section and the remainder of the movement are each framed. In the first case, the frame consists of two musical metaphors. Shortly before quoting Knecht's report of the troubling debate, the chronicler announces "the psychological history of the friendship-enmity between Joseph and Plinio" by calling it "a sonata movement on two themes" (p. 93). Having documented Plinio's farewell words, Ferromonte comments that "to me, as a musician, this confession of Plinio [...] was like a musical experience. The contrast of world and Mind, of Plinio and Joseph, had before my eyes been transfigured from the conflict of two irreconcilable principles into a double concerto" (p. 108). The block comprising sections 2-5 is preceded by both men's expressions of gratitude for having been able to share aspects of their different lives, and rounded up by a true experience of spiritual sharing when Knecht plays for his friend a movement from a Purcell sonata.

The third movement of the assumed *ludus*, Joseph Knecht's circular to the Castalian directorate, is framed by preliminary remarks and a coda as was the chronicler's introduction in movement 1. Its two themes are those from movement 1. The World of the Mind (WoM), considered here with an emphasis on the dangers it ignores and a warning of the decline it must face if it does not readjust, serves as theme 1. It is set against a reminder of the feuilletonistic age (AoF), evoked here with the intent of learning from a past gone awry. The two themes are corroborated by the ideal of service as motif 1, the need for historical awareness as motif 2, and the problem of non-utilitarian luxury, exemplified in the glass bead game (GBG) as motif 3.

FIGURE 12: Thematic material and structure in Knecht's circular

preliminary remarks with m1 (SERVICE)	
(1) exposition and development of theme 1 (WoM) with assistance from m1 (SERVICE), m2 (HISTORY), and m3 (glass bead game)	pp. 346-351
(2) brief confrontation of the themes (WoM / AoF) with frequent and strong emphasis of HISTORY	pp. 351-352
(3) exposition and development of theme 2 (AoF) with reminiscences of HISTORY and GBG	pp. 352-355
(4) restatement of theme 1 (WoM) with m2 (HISTORY) and m3 (GBG)	pp. 356-357
(5) two flawed and one desirable suggestions for a synthesis of the two themes, the latter with m1 (SERVICE), m2 (HISTORY), and strong emphasis on the now highly problematized m3 (GBG)	pp. 358-363
coda with quotation from Father Jacobus (HISTORY)	

Movement 3 thus reworks the first movement's themes in a new light. The relationship between Castalia (with its glass bead game) and the depraved 20th century is inverted, with regard both to their position in the musical structure and to the value attributed to each. In the introduction, the chronicler had presented the Age of the Feuilleton as a past about which one laughs and marvels as about a children's world—a past from which Castalia has risen to shine forever. In Knecht's circular, by contrast, the World of the Mind appears as imperiled if not already declining, the glass bead game as an extravagance, and the feuilletonistic era as an example to be heeded as a warning.

Moreover, movement 3 harkens back to movement 1 through explicit allusion: the historian Plinius Ziegenhalss is mentioned again, and the first movement's three quotations are matched by two shorter quotations in Knecht's missive. While the chronicler's use of extant texts is unambiguous in its aim to corroborate his argument by enlisting the wisdom of the great epochs of world culture (Chinese antiquity, the European Middle Ages, and Castalian modernity), Knecht's borrowings are astute and dialectic in that they juxtapose an example of academic prostitution with a voice of intellectual responsibility. A "university professor in the Republic of the Massagetes" (Hesse's term for Nazi Germany, hence part of theme 2) is quoted as saying: "Not the faculty but His Excellency the General can properly determine the sum of two and two" (p. 354), whereas the late-19th-century historian Jacob Burckhardt's reflections on how culture can and must be preserved in times of "terror and deepest misery" (p. 363) are called for support of Knecht's concern about the crisis faced by the World of the Mind (theme 1).

In the fourth movement—the Directorate's reply to the circular—all direct responses to Knecht's analysis are relegated to the thematic material's secondary level: "historical awareness" is now motif 1 and "service" appears as motif 2. Both appear as little more than polite acknowledgments of aspects the Castalian bureaucrats consider quite irrelevant to the question at hand. The contrasting themes, which address Knecht's suggestions for improving the relationship between the World of the Mind and the world-at-large, are those of the Knecht-Designori conversation, taken up with inverted value. In other words, movement 4 revisits the movement-2 themes just as movement 3 takes up those from movement 1. In the principal theme of movement 4, the Directorate expresses its praise for Knecht's analysis, couched in a tone that is condescendingly jovial and grating in its excessive show of attempted fairness ("sham harmony"). In the secondary theme, all his conclusions are rejected on the grounds that a closer Castalian involvement with mundane issues is deemed neither feasible nor desirable ("distance"). The final component, motif 3, addresses the submissive relationship the individual ought to retain vis-à-vis the hierarchy; Hesse thereby closes the circle to the chronicler's praise of the Castalians' "integration into the generality" in the first ludic movement. This motif manifests itself in two complementary aspects: Knecht's self-willed violation of Castalian obedience is chastised as "frivolous behavior," "tasteless assault upon the tranquility and the imagination of [...] colleagues" etc. (motif 3a), and counterbalanced with a veritable battery of densely spaced verbal bows to "the majority" (motif 3b).

Movement 4 is much shorter than the other three but once again laid out in five segments. Its design follows that of sonata allegro movements, of the kind often found in the repertoire of the pre-classical and early classical periods.

FIGURE 13: Thematic material and structure in the Directorate's reply

(1) theme 1 (SHAM HARMONY) with mention of m1 (HISTORY), m2 (SERVICE), m3a (SELF-WILL)	pp. 364
(2) theme 2 (distance) with m1 (HISTORY) and m3b (MAJORITY)	pp. 365-366
(3) restatement of theme 1 (with belabored emphasis on motif 3a)	pp. 366-367
(4) development (a furious contest of motifs 3a and 3b)	pp. 367-368
(5) recapitulation of theme 1 and motif 2	pp. 368

The temporal progression in the four movements of what I propose to read as a discursive glass bead game is worth a brief comment. Movements 1, 2+3, and 4 track the history of the relationship between the World of the Mind and the world-at-large backwards. The 25th-century chronicler's idealizing view emerges as having been made possible by a far-sighted 23rd-century *magister ludi*'s willingness to act on his insight into the dangers of intellectual pride and class arrogance. This assessment is followed in the story by a display of precisely the attitude that caused Knecht's concerns—an attitude that has most likely served as accepted wisdom for quite a while and thus points back to even earlier times.

In terms of the four movements' treatment of the themes and their gradual approach toward synthesis, there is mirror inversion in all dimensions.

- The antithesis in the themes of movements 1 + 3 and 2 + 4 has already been highlighted: The World of the Mind, posited as a glorious achievement, comes to look sterile and conceited, while the Age of the Feuilleton, initially derided, appears as a valuable historical lesson. Distance between the worlds of mind and nature, perceived as deplorable by the debating partners, is declared inevitable and desirable by the Castalian Board of Directors, while their condescending sermon of mock praise turns Knecht's plea for harmony into sham harmony.
- This *explicit* antithesis is complemented by *implicit* antithesis in movements 1 + 2 and 3 + 4 . Whereas the chronicler in his introduction barely recognizes and never truly acknowledges the problem of Castalia's estrangement from the larger world, the debating partners make distance versus harmony their leading juxtaposition. This awareness is taken to the next level when a Castalian *magister* no longer merely analyzes and laments the rift but embarks on exemplary action. Improved harmony seems within reach. Yet such hope is thwarted in the Board's reply, when the directors disregard Knecht's premonition and reject the solutions he suggests, arguing that closer involvement with the world's concerns is not in Castalia's interest.

As the first four movements show, this discursive glass bead game satisfies the chronicler's category of a "formal" game: it is a model of clarity and transparency. Moreover, from the musical point of view, the succession of structures seems to trace the early development of sonata allegro form.

Nonetheless, the task of synthesizing the contrasting themes has not yet been achieved. A further act will be needed in this symbolic drama. This will bring the number of movements up to FIVE, Hesse's magical number. While the conclusion may at this point be eagerly awaited as the desirable final step within the development of this game, its realization in the interwoven threads of the double-stranded story has to wait. It is contingent upon the preparatory move toward synthesis undertaken by the "soul" whose developmental history Hesse traces in the other strand. Joseph Knecht may have composed his poems while in Waldzell and may remember them immediately prior to his resignation from Castalia. The reader, however, will only be ready for this conclusion after Magister Knecht has reached the "beautiful bridge" (Belpunt) and his life has consummated harmony in the interpersonal sphere.

With this recognition, we may step back and allow our gaze to take in the entirety of the two intertwined strands. It seems to me crucial that we imagine their 5 + 5 segments not as a simple alternation in linear fashion, but rather as a kind of double helix.

FIGURE 14: A glass bead game intertwined with a "story of a soul"

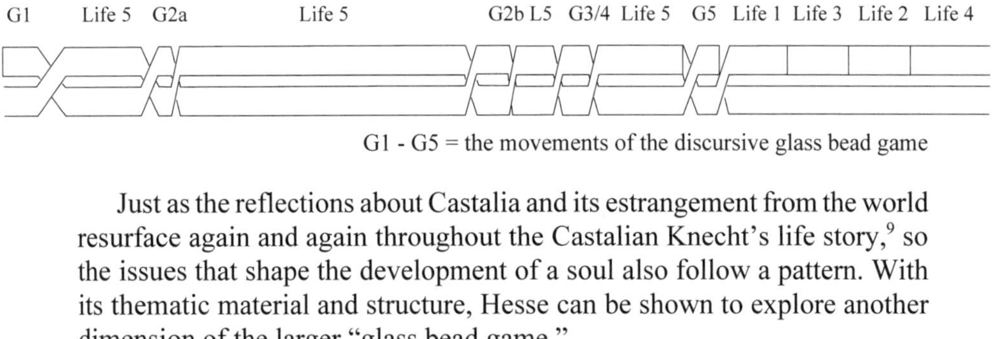

G1 - G5 = the movements of the discursive glass bead game

Just as the reflections about Castalia and its estrangement from the world resurface again and again throughout the Castalian Knecht's life story,[9] so the issues that shape the development of a soul also follow a pattern. With its thematic material and structure, Hesse can be shown to explore another dimension of the larger "glass bead game."

[9]See particularly the extensive deliberations offered by the chronicler (C), Knecht (K), and the Music Master (MM): p. 61/line 20 to p. 66/line 16 = C, 69/4-71/5 = K, 71/22-72/12 = K, 73/23-75/3 = MM, 76/10-77/5 = K, 81/13-83/20 = MM, 85/4-12 = MM; 96/16-97/27 = K, 100/1-35 = K, 101/18-104/34 = MM, 109/20-24 = K; 110/35-113/27 = C, 118/8-121/10 = K, 121/24-122/5 = K; 133/27-134/17 = C, 136/25-137/2 = K; 149/35-150/27 = C, 197/3-37 = C; 204/3-205/10 = C, 208/29-210/5 = C, 233/15-237/33 = K; 264/13-20 = K, 264/21-272/33 + 274/17-275/17 + 275/22-277/6 = C, 278/32-279/21 = K, altogether another 36 pages.

Part III
GEOMETRY

PLATES 3a - e: The Platonic Solids and the cosmic elements

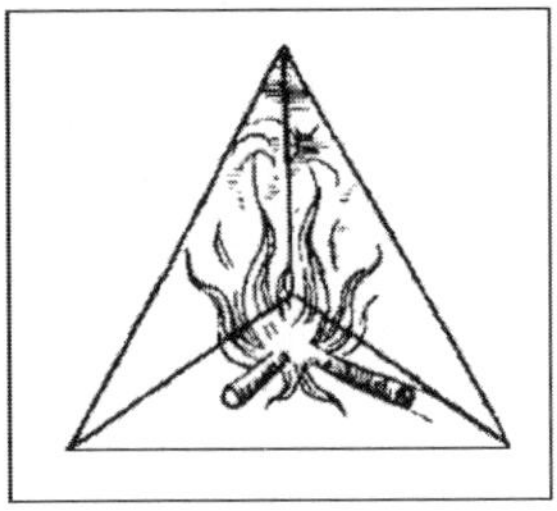

tetrahedron
(fire)

cube
(earth)

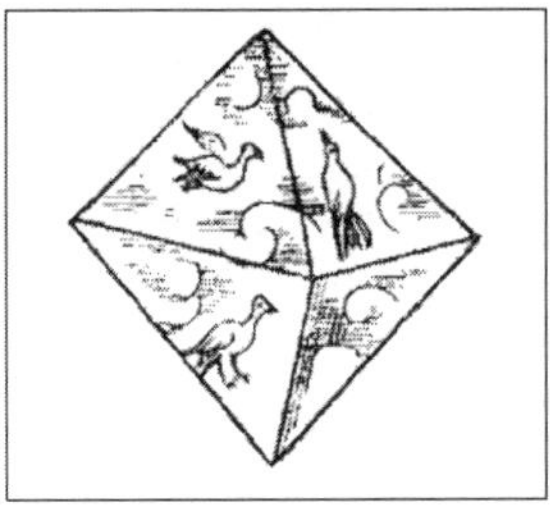

octahedron
(air)

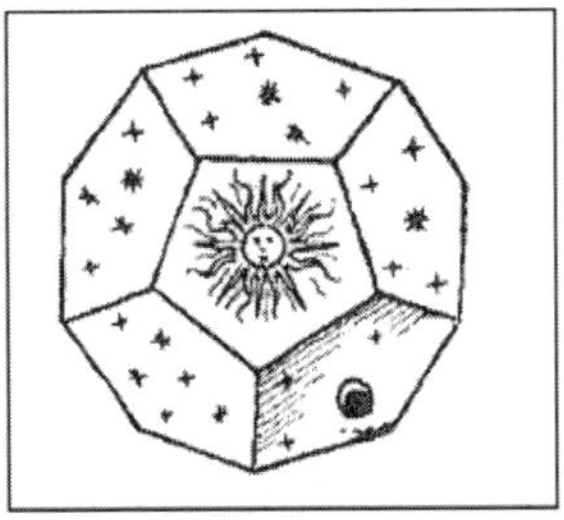

dodecahedron
(sky)

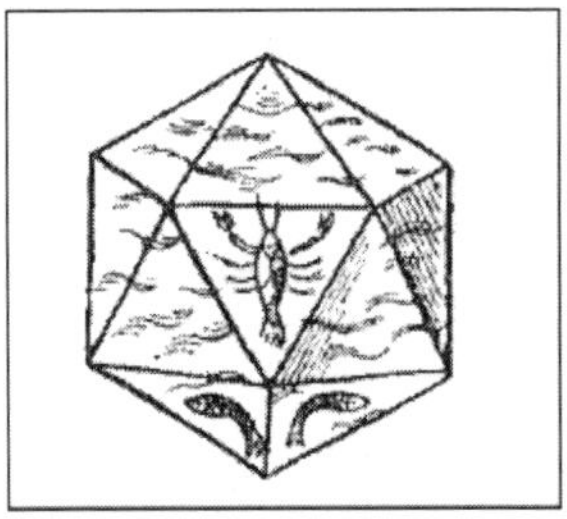

icosahedron
(water)

Ideal Figures and Bodies

While more mystically inclined scholars of universal harmony believe that the *lambdoma* matrix takes pride of place in representing something like the Divine Geometer's plan for all components on earth,[1] practical applications for similar claims are usually based on the *tetraktys*. For geometers, its ten dots invite myriad connections. 42 lines, linking two dots each (fig. 14), build many of the figures that Euclid, the most famous geometer in the Pythagorean tradition, discusses in his *Elements*. Some of them are obvious, like the ten equilateral triangles—nine of them small, among which six point up and three point down, and a large one enclosing the others. Not so obvious are the many right-angled triangles, famous for the theorem attributed to Pythagoras ($a^2 + b^2 = c^2$). 36 of these, all of equal size, can be drawn into the *tetraktys* (fig. 15 separates them into 9 groups of 4).

Another figure to appear in the *tetraktys* is the helicon: see its square body and its two diagonal strings, segmented at the ratio 6 : 8 : 9 : 12.

FIGURE 15: A geometric exploration of the *tetraktys*

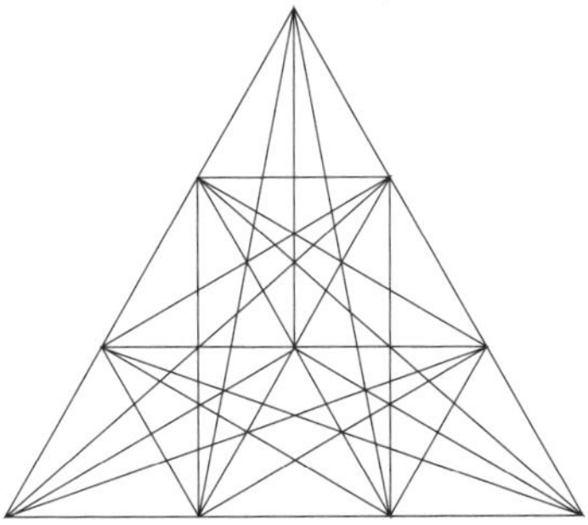

Figure 16: Right-angled triangles in the *tetraktys*

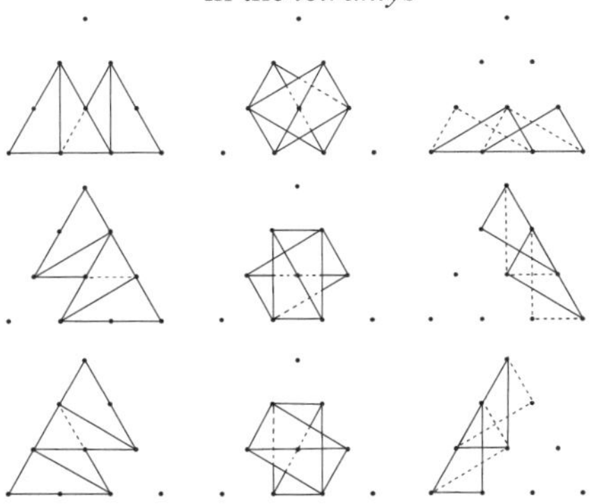

FIGURE 17: The helicon in the *tetraktys*

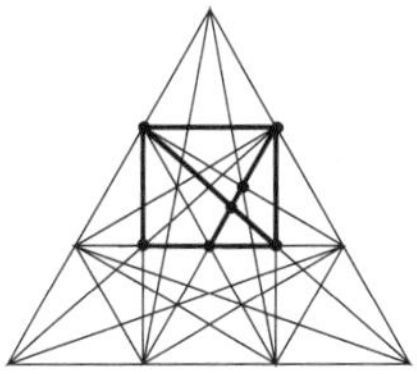

[1]Rudolf Haase, commenting on widely shared albeit probably esoteric convictions among thinkers in this tradition, believes that "the *lambdoma* matrix could have the function of a cathedral builder's canon for the earth and its beings." *Natur, Geist, Seele: Harmonik und Metaphysik des quadratischen und des runden Lambdoma* (Vienna: Braumüller, 1985), p. 54.

FIGURE 18: Trigram, hexagon, and hexagram

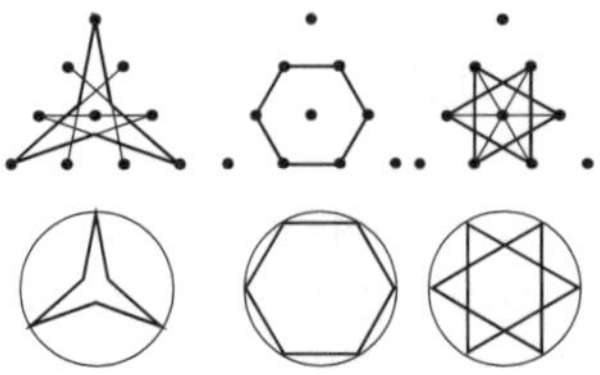

Finally, there are three further regular figures that, though more often seen inscribed into a circle, also form part of the geometric web in the *tetraktys*: the trigram, the hexagon, and the hexagram. Owing to their connection to both circle and *tetraktys*, they create a link between, as it were, the esoteric and the exoteric representatives of the most perfect shape in geometry.

The regular figures with four or five angles have no relationship with the *tetraktys*. As if to compensate, they have other proportional properties. The square allows for a process called the quadrature. If a given square is filled with further squares in such a way that all corners bisect the sides of the next larger figure, then their sizes decrease proportionally: every square is half the size ("an octave") of the next larger one with parallel sides, while each tilted one relates to the straight one around it as $\frac{1}{2}$ a x $\sqrt{2}$, a ratio close to that of the tritone ("half an octave").

FIG. 19: Quadrature and its proportions

Five-pointed figures are seen as set apart. They are considered mystical in most traditions, owing to the symbolic value attributed to the number 5. The pentragram particularly was also deemed magical: it is the simplest star shape that can be drawn with a single line (for this reason it is sometimes called the Endless Knot) and with no other reference than the angles on the pentagon's circumference. Pentagrams have been found on potsherds at Ur dated to ca. 3500 B.C.E. In later periods of Mesopotamian art, pentagrams appear in royal inscriptions, symbolizing imperial power. The Hebrews associated the number five and the symbols based on it with Truth and the five books of the Pentateuch, just as the Christians later linked them to the five wounds of Christ and the Muslims know five pillars of faith and five daily times of prayer. In a similar spirit, the Pythagoreans regarded the pentragram as an emblem of perfection. In the medieval legend of Sir Gawain and the Green Knight, the pentagram was Sir Gawain's glyph, inscribed in gold on his shield, symbolizing the five knightly virtues: generosity, courtesy, chastity, chivalry and piety.

In a slightly different development, medieval thinkers and artists who perceived the concept of the microcosmic world of Man as analogous to the macrocosm of the greater universe of spirit and elemental matter, used the pentagram as a symbol for humans within their microcosm, implying an analogy to the macrocosmic whole. Kepler's predecessor at Prague and short-term mentor, Tycho Brahe, reproduces in his *Calendarium Naturale Magicum Perpetuum* (1582) a pentagram with a human body inscribed into it and the Hebrew letters for YHSVH associated with the elements. Other illustrations of the period, by Robert Fludd and Leonardo da Vinci, also show geometric relationships of man to the universe. Both in such inscribed representations and in its pure geometric form, the pentagram symbolizes (microcosmically) the relationship of the human head to the four limbs and hence (macrocosmically) that of the spirit to the four material elements: earth, water, air, and fire.

FIGURE 20: The Pentagram and the golden section

Returning to geometry and harmonious proportions, the pentagram evokes a ratio not accessible through any of the other figures: the golden section. The continuous line crosses each of the five-pointed star's sides at two points, each time bisecting it so that the shorter segment relates to the longer one as the longer does to the whole. Moreover, the ratio of a central part of any of the pentagram's lines to any of the outer parts is in the golden proportion, as is the ratio of the height above the horizontal bar to that below.

Diminution by the golden ratio allows to fit figures into one another: aligning proportionately smaller ones to the longer segment of the next larger pentagram's horizontal results in a harmonious sequence of diminishing figures. Or, building a square over this segment and a "golden rectangle" beside it, partitioning another square from the rectangle and thereby inscribing a smaller version of the dual figure into it, etc., produces the geometric frame for a famous figure of golden ratios: the nautilus spiral.

FIGURE 21: Proportional play with the golden section

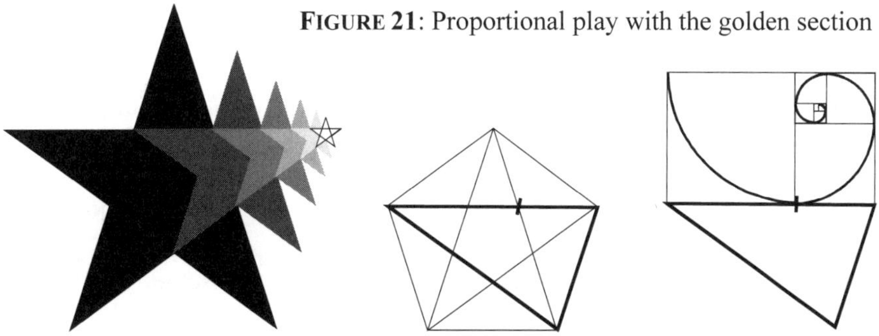

Finally for harmony in geometry, there are the regular polyhedra: solids bounded by regular polygons, or: three-dimensional figures whose faces, edges, and vertices are all equal. In nature they occur in gem crystals and other minerals. Architects use them in the understanding that humans have an innate inclination to bodies and spaces built to perfect proportions. Only five regular polyhedra are possible: those of 4, 6, 8, 12, and 20 faces. They are called tetrahedron, cube, octahedron, dodecahedron, and icosahedron and have come to be known as the Platonic solids ever since Plato associated them with the primal elements of which, as Empedocles (ca. 493-433 B.C.E.) had observed, all things are composed. Plato assigned

- the tetrahedron or pyramid, which consists of four regular triangles, to fire (as he explains in the *Timaeus*: "The figure that has the fewest faces must in nature be the most penetrating"),
- the octahedron, constructed from eight regular triangles, to air,
- the icosahedron, built from 20 regular triangles, to water,
- and the cube, made up of six squares, to earth ("for it is the most immobile of bodies and retentive of shape").

This exhausts the four elements; the fifth solid had to be associated with something else. For the dodecahedron, whose twelve faces are pentagons and which is thereby in many respects different from the other solids, Plato thought of the cosmos, stating that the demiurge used it for "embroidering the constellations on the whole of heaven."

Like the pentagons that build its faces, the dodecahedron is intimately related to the golden section. In addition to the properties of the plane figure, its surface area and its volume of unit edge length are simple functions of that ratio. At the same time, the number of its faces, 12, is associated with the months in a year, the hours in a day, the zodiac—cosmically defined features.

In Book I of *Harmonices mundi*, Kepler discusses Euclidean geometry in general and the *proportio divina* or *sectio proportionalis* in particular. He returns to it later when, in chapter 15 of Book III, he develops it from what we now call the Fibonacci series. Kepler reminds us that Euclid, tracing the gradual approach of Fibonacci's ratios to the mean, argued that the most common human application of the golden section occurs in the form of the proportion 5 : 8. In other words: the Pythagorean geometer was thinking of *musical* harmony rather than of the mathematical mean.

Of the many occurrences of these proportions in nature's geometry, I want to mention three: the inverse relationship between sight and hearing with regard to perspective, the ratios defining the shapes of crystals, and the laws governing motion and structure in atoms, molecules, and biological rhythms.

Leonardo da Vinci discovered that there are significant correspondences between the ways in which the human eye and ear perceive impressions. To give an example, there is a direct—albeit inverse—relationship between perspective and the overtone series (or, as the analogy has been described, between "visual and aural perspective").[2] Objects placed at regular distances from one another are seen in a foreshortening that corresponds with one wing of the lambdoma: If the distance at which the human eye perceives the first object is taken as a measure ("1"), the subsequent object will visually appear as half as far from the first, the third, as $\frac{1}{3}$ as far from the second, and so on through $\frac{1}{4}$, $\frac{1}{5}$, $\frac{1}{6}$, etc. This is inversely analogous to the way in which the increase in the speed of vibrations is heard. In other words:

objects placed at *equal distances* in reality are seen *perspectivally*
(trees T . . . . . T . . . . . T . . . . . T . . . . . T . . . . . T . . . . . T . . . . .
appear as T . . . . T . . . . . T . . . T . . T . T T)

tones heard as *equal distances* are *perspectivally* spaced in reality
(octaves A_1 . . . . A_2 . . . . A_3 . . . . A_4 . . . . A_5 . . . . A_6 . . . .
sound at 55 . 110 . . 220 . . . . 440 . . . . 880 . . . . 1760 Hz)

Ever since Kepler had begun to prove what Pythagoras had conjectured ca. 2160 years before him—that the heavens, the world of consonant sound, and the human condition are determined by identical norms of proportionality—scientists have speculated that the same must hold true for other components of growing nature: that the principle of harmonic ratios must also inform the structure of atoms, molecules, and ultimately, of all living organisms. Crystallography became the first field in which large-scale investigations were conducted. In 1809, the German crystallographer Christian Samuel Weiss showed that the polyhedral structures of crystals, particularly in the proportions between their edges and planes, could be represented by simple indices. In 1839, William Hallowes Miller developed these indices into a crystal notation system. Building on this, Victor Goldschmidt (1853-1933) undertook wide-ranging studies proving that the intercepts at which different faces of a given crystal cut each axis are always related to one another in the form of simple whole-number multiples. With the help of illustrations in the *lambdoma* matrix, he showed that the harmonic ratio is essential in the world of crystals and their growth, and that their proportions are directly analogous to musical intervals. In the most influential of his many books, *On Harmony and Complication* (originally published 1901 in Berlin), Goldschmidt demonstrates the proportions of the entire musical scale at the microscopic level in the relative dimensions of crystals.

[2]Leonardo da Vinci, *Trattato della pittura* (Rome, 1651), Book IV, p. 581-582; and Rudolf Haase, *Geschichte des harmonikalen Pythagoreismus* (Vienna: Lafite, 1969), pp. 76-77.

Encouraged by Goldschmidt, corresponding research into similar properties in the arrangement of matter in general blossomed in the mid-20th century. The Nobel prize-winning physicist Nils Bohr found that, in the first four shells of the hydrogen atom, the single electron moved at exactly proportional speeds: at 2160 kilohertz per second in the first shell; at 1080, 720, and 540 kHz/s respectively in the second, third, and fourth shells. In musical terms, the electron in the second shell creates the octave below that in the first shell; in the third shell, we hear the fifth below that lower octave, and in the fourth shell, the fourth below the fifth—a natural occurrence of the *under*tone series. The mathematician H.E. Huntley, discussing the golden section in his book *The Divine Proportion*, describes the energy exchange between several levels of electron shells and suggests that this exchange, rising and falling in different amounts from one shell to another, occurs in a pattern that follows the ratio of Fibonacci numbers. Huntley affirms the proportion of the musical minor sixth, or the ratio 5 : 8, in the quantum leaps between energy levels of simple atoms of a hydrogen gas.[3] C.N. Banwell, writing in a highschool physics text book, describes spectral analysis of energy levels in simple electron organization using terms like "fundamental," "overtones," "combination frequencies" etc.[4] Donald H. Andrews, chemistry professor at Johns Hopkins University in the mid-20th century, writes enthusiastically about the musicality of atoms.[5]

Even botanists find that not only the components of blossoms, but also spiral growth patterns show harmonic ratios: the number of leaves between two vertically aligned sprouts and the number of revolutions traced relate to one another either as musical intervals (1 : 2, 1 : 3, 2 : 5, 3 : 8, etc.) or in the Fibonacci ratio.[6] Last but by no means least, the medical researcher Gunther Hildebrandt confirms the same proportions for human biological rhythms, diagnosing pending disease by noting deviations from the ratio.[7]

The world, when healthy—i.e., in tune with numerical ideals—is alive with consonance.

[3]H. E. Huntley: *The Divine Proportion: A Study in Mathematical Beauty* (New York, Dover Publications, 1970), p. 156.

[4]C.N. Banwell, *Fundamentals of Molecular Spectroscopy* (London, New York: McGraw-Hill, 1966).

[5]Donald Hatch Andrews, *The Symphony of Life* (Lee's Summit, MO : Unity Books, 1967), and *Consonant Form in Music and Art* (Dallas: Dallas Independent School District, 1963).

[6]Hans Kayser, *Harmonia Plantarum* (Basel: B. Schwabe, 1943).

[7]Gunther Hildebrandt, *Chronobiology & Chronomedicine: Basic Research and Applications* (Frankfurt : P. Lang : 1993) and *Biological Rhythms in Humans and their Counterparts in Music* (forthcoming).

Kepler's *Somnium* and Hindemith's Rondos

Act III of Hindemith's opera centers on a part of Kepler's creative imagination that combines the scientist's passion for astronomy, a portrait of Kepler's mother, and the composer's interest in ideal figures and bodies.

Abandoned by a husband who sought the adventures of a mercenary life, the historical mother Kepler was harsh and haughty. She survived thanks to a special gift: her intimate knowledge of the medicinal effects of plants and herbs, handed down through generations of wise women in her family. But her abrasive behavior provoked her neighbors, who eventually used her uncanny healing abilities as an excuse for getting at her by denouncing her as a witch. After years of written charges and defenses (the latter from her son Johannes's pen), she was arrested in 1620 and, but for the dispense Kepler obtained from the High Court, would almost certainly have been executed.

Beyond filial loyalty, Kepler had strong reasons to do everything in his power to help his mother, since he may have involuntarily contributed to her plight. As a student in Tübingen, Kepler had written a fictional story about a voyage to the moon.[8] Drawing on his fascination with astronomy and meteorology, his extraordinary talent for spatial imagination, and a fertile poetic mind, Kepler's *Somnium* describes in great detail not only the moon's geography and climate, but also the living conditions of moon dwellers as well as the appearance of the starry sky from their perspective. This essay—which Kepler, following a long tradition, couched as a dream—has been called a path-breaking predecessor of science fiction.[9] Although it was printed only posthumously, it is assumed to have circulated from about 1608 onward in several handwritten copies.

In a later revision, Kepler framed the "travelogue" with a story about an Icelandic folk herbalist and her son, an aspiring scholar who leaves home to train as an astronomer. Returning from his studies, the young scientist urges his mother to reveal the source of her unusual knowledge, and learns that she has long been receiving instructions from a demon. It is this demon, now purposefully conjured by mother and son together, who then describes the voyage to the moon that was the initial cell of Kepler's imaginative text. Only much later did Kepler understand what conclusions this tale could invite in the heyday of witch trials, especially since he had portrayed the fictitious astronomer's mother and alleged witch with the character and demeanor everybody resented in his mother.

[8]It was eventually published under the title *Somnium seu opus posthumum de astronomia Lunari.* For an English translation see Edward Rosen, *Kepler's* Somnium*: The Dream, or Posthumous Work on Lunar Astronomy* (Madison, WI: University of Wisconsin Press, 1967).

[9]Mechthild Lemcke, *Johannes Kepler* (Reinbeck bei Hamburg: Rowohlt, 1995), p. 96.

In act III of *Die Harmonie der Welt*, Hindemith combines two scenes that must be read as creative extrapolations of the two components comprised in Kepler's "Moon Dream" essay: the interest in the moon, along with that in the change of perspective a voyage to it would entail, and the relationship between a cranky folk herbalist (who may or may not be a witch) and her astronomer son. The link may be credible enough from the content of the scenes Hindemith finally set to music. But there is an even stronger evidence for the connection I am suggesting. Alfred Rubeli, one of the composer's students in Zurich at the time when Hindemith worked on the libretto and later the executor of his estate, tells of an earlier, much more extensive version of the text, in which act III began in the following way:

> (From the depth of darkness, light spreads across the upper half of the stage. It represents the surface of the moon with mountain peaks, craters, and maria. In the background one sees the sphere of the earth turning slowly. Half hidden behind passing clouds one distinguishes parts of Europe and Africa. Kepler is seen sitting to one side, pondering.)
>
> Kepler: This is what I'll have to describe one day:
> how the world appears when seen from the moon,
> and the earth. There is its ball, spinning
> some thirty times within one lunar day.
> There is Africa, a head that bends down
> to kiss a girl that beckons a kitten:
> Northland, as it leaps toward Europe.
>
> (The moon people have assembled. They gather around Kepler, singing and dancing a lunar *pavane*.)
>
> Kepler: This is what the Lunarites are used to seeing.
> A learned fairy-tale, the dream about the moon.[10]

The protagonist's words about "what I'll have to describe one day: how the world appears when seen from the moon" as well as his explicit reference

[10]Translated from Alfred Rubeli, "Johannes Keplers Harmonik in Paul Hindemiths Oper *Die Harmonie der Welt*," in Rudolf Haase, ed., *Kepler Symposion. Zu Johannes Keplers 350. Todestag*, 25.-28. September 1980 im Rahmen des Internationalen Brucknerfestes (Linz: Linzer Veranstaltungsgesellschaft, 1981), pp. 107-116 [108]. Winrich Hopp also mentions this draft in his essay "Von der harmonikalen Fehlgründung lebenspraktischen Glücks: Zur Sinfonie und Oper *Die Harmonie der Welt*," *Hindemith-Jahrbuch* 2002/xxxi: 19-40 [34-36]. Note that the text quoted by Rubeli differs from the early draft of the libretto published and discussed by Andres Briner in "Die erste Textfassung von Paul Hindemiths Oper 'Die Harmonie der Welt'," in Carl Dahlhaus, ed., *Festschrift für einen Verleger. Ludwig Strecker zum 90. Geburtstag* (Mainz: B. Schott's Söhne, 1973), pp. 203-241.

to the "dream about the moon" leave little doubt that Hindemith was familiar even with this lesser-known piece of Kepler's writings, and took inspiration from it when he sketched the central act of his opera.

In the final libretto, in which many colorful scenes fell by the wayside in the interest of brevity and clarity, Hindemith abandoned the idea of staging the astronomer's dream-like stay on the moon. Instead he coupled the scene addressing mother Kepler's alleged witchcraft with one exposing various concepts about the moon. While mother Kepler, having fled from the accusations at home, visits her son's house in Linz, the various family members are shown musing about the moon, attaching very different significance to its impact on their souls and life goals. Back in Württemberg, mother Kepler is brought to court on charges of practicing witchcraft. During the trial, the witnesses, the listeners, and the prosecutor sound more superstitious than the accused. Even mother Kepler's eventual acquittal, owed to her famous son's intervention and connections to higher-ranking officials, relies on "magic" rather than on evidence. Once set free, however, mother Kepler discloses the full extent of her fantastic ideas and ambitions. The two scenes, which seem rather different in their dramatic presentation, are thus intimately related. What is more, both are musically set as unusual variants of the rondo form.

So what is the manifold significance of the moon, explored in the first rondo? For naive souls like Kepler's little daughter, the moon is humankind's affectionate if unknowable friend, inhabited by "the little man in the moon" whose solitude is to be pitied; for superstitious people like mother Kepler, the moon is an accomplice and facilitator of all manner of dubious acts; for conscientious people like Kepler's second wife Susanna, moonlight defines a time of reflection and self-questioning set apart from the busy sunlit hours of everyday life; for a scientist like Kepler, the moon is an object of investigation, perhaps a means toward a growing understanding of the eternal laws of the cosmos. A surprising fifth dimension is added by the chorus of "voices from the moon," who describe the moon's primary quality as being only and fully herself. They assert the moon's intrinsic lack of emotions (whether compassionate or vengeful) and debunk all myths humans have invented regarding alleged lunar interventions in human endeavors and fates. The many sides from which the moon is seen here function as an antidote of sorts to the superstitions that dominate the subsequent trial scene.

The "moon rondo" is held together by a refrain that juxtaposes a pastiche children's song addressed to the "Little man in the moon" with choral statements by the lunar voices about the "world without clemency, airless, cloudless, rainless." The folk-song component in the first half of each refrain, sung by Kepler's daughter to woodwind accompaniment, modulates from its tonal center on B to D (cf. rehearsal cues [17-17B]). The choral responses in the refrain's second half lead from D back to B, thus symbolically complementing the child's view of the moon.

EXAMPLE 10: Song to the "little man in the moon" and reply from the moon

The first episode, which features the unexpected meeting between two very different women—Kepler's wife and mother—falls into three sections: a brief dialogue, mother Kepler's tale of the tribulations that caused her to flee, and Susanna's welcoming words. Of the three segments in the second episode, the first two also juxtapose the two women. In separate chambers on a split stage, they reflect on their feelings of discomfort with one another and then seek confirmation for their natural inclinations in Biblical passages, which they recite in the style of scripture chanted in church: Susanna ponders verses from Laments 3 that extol self-abasement and generosity toward others; mother Kepler identifies with Isaiah who, in chapter 42 of his book, describes how he learned that he was chosen to enlighten others and decided that he would now no longer hold his tongue but cry, destroy, and devour. While the audience cannot yet understand for what mother Kepler may think she has been chosen, her resolution to "destroy and devour" without restraint is chilling.

In the second episode's final component, Kepler's daughter expresses her concern for the continued well-being of the man in the moon and asks what her father thinks it is like "up there." Kepler answers absent-mindedly with descriptions of what he has seen through his telescope, mentioning jagged mountains that cast their shadows to different sides depending on the relative position of moon and sun. Absorbed in his thoughts, he proceeds to expound his first law of planetary motion, concerning the elliptical form of all orbits. When the girl asks anxiously how the man in the moon takes care of the lunar mountains each time the crescent narrows, it becomes obvious to what degree the father is out of touch with his daughter's mental horizon.

Instead of responding to her concern in a way that she could understand, Kepler now launches a lengthy monologue in the form of a hymn.

This hymn is based on a recurring thematic phrase. Its twelve statements —six each in two analogous halves—all sound in the lowest register beneath a largely homophonic texture, as is typical in a passacaglia. At the same time and atypical of a passacaglia, the statements wander through different keys as would the subject entries in a fugue, and are each challenged by a genuine counterpoint (a bass-register instrument whenever Kepler's vocal line takes up the subject, and vice versa). In the hymn's first half, Kepler—still standing next to his baffled young daughter—stresses the importance of "the formula" and adds succinct definitions of his second and third planetary laws (singing "where equal areas fit into the elliptical orbits in equally long times" and "the revolution of two celestial bodies in squared and cubed proportion revealed through God's wisdom").The meter in the thematic phrase alternates regularly between $\frac{2}{2}$ and $\frac{3}{2}$ time, as though Hindemith wanted to remind listeners that it was Kepler's search for simple proportions that inspired his insights.

EXAMPLE 11: Kepler reflects in the form of a hymn [18]

Wo nichts Greif- - ba-res, wo die For-mel wohnt,
(Where there is nothing tangible, where the formula dwells)

The varied repetition in the hymn's second half substitutes Kepler's all-too-learned monologue with a vocal quartet. Mother Kepler and Susanna return to expressing their troubled feelings toward one another, again with biblical quotations: Susanna reads in Job 15 that self-righteousness displeases God and vows to be humbler, whereas mother Kepler finds confirmation in Revelation 3 that God scorns lukewarm behavior. Meanwhile, Kepler thanks God (in Latin words taken from one of the famous prayers at the conclusion of *Harmonices mundi*) for having been allowed to make manifest the glory of His Creation—the laws of universal consonance, and his daughter exclaims wistfully that, after all the learned descriptions she has heard, she feels homesick for her dear old moon.

These words are the cue for the transition to the final stanza of the children's song about the little man in the moon. The final response by the voices from the moon, heard as the curtain has already fallen over the scene in Kepler's house, reiterates that humans' insincere, indecisive, and selfish behavior is not caused by, but rather exposed in, the moon's icy light.

Rondos, as is well known, originated as rustic dances. Their refrains are typically simple enough in melody, rhythm, and structure to be memorable for the crowd of untrained dancers asked to skip around a circle. Episodes, inserting show dances by featured couples, may be more ambitious in all respects but are still to be danced. A hymn inserted into a rondo—and what is more, one composed as a hybrid of passacaglia and fugue—is about as fitting as would be a religious sermon preached at the height of a cheerful garden party.

The "trial rondo," which begins at cue [19] and fills the remainder of act III, is composed as an expanded analog to the moon rondo. Almost exactly twice as long in the score (652 : 328 measures), the second rondo enhances several of the traits found in the first, exaggerating them and thus matching the weirdness of the court proceedings. Other features recur oddly displaced —another procedure that is symbolic of the scene's content. What seems incongruous in the moon rondo is exacerbated in the eccentricity of the trial rondo: here it is the refrain that is cast as a strictly polyphonic texture. In fact, the recurring phrases are not, as in normal rondos, fundamentally identical, but rather appear as something like severed segments of an ongoing fugue.

Several features betray the function of this fugal subject as an emblem of mother Kepler's deranged fantasies long before her words do. Its twenty-four entries, complemented by numerous detached incipits and tails, point to the obsessive nature of the old woman's thoughts. The fact that none of the vocalists ever takes up or joins the subject's contour suggests that mother Kepler's plan exists in a realm not accessible even to the witnesses and prosecutors, much as they are prepared to find her guilty of witchcraft. (When she finally does divulge her ploy, she uses different musical material —what I am calling the "revelation theme.")

The tonal ambiguity in the refrain's "fugal subject," whose intervals suggests three cross-related chords in succession (F♯ major, F♯ minor, and F major), seems like an almost too literal symbol of schizophrenia; the obstinate hemiolic syncopations (**1** 2 **3** | 1 **2** 3 etc.) hint at a mental fixation. For its last four entries, following Kepler's rejection of the "lucrative collaboration" with his mother, the fugal subject sounds yet different: half as slow as before and with a severely distorted rhythm. This change confirms what the audience may have started guessing at this point: that this woman's courage and stamina depend entirely on her madcap dream. Once her son has shattered all hopes for its realization, she is likely to lose her will to live, if not her mind.

EXAMPLE 12: Mother Kepler's secret hope ([19] and [21 L])

If the non-vocal fugal subject stands for the never-verbalized scheme concocted in a mind that has lost its sense of reality, the distribution of its entries over the course of the rondo—with 7, 5, 3, and 1 in refrains 1-4, transformed and fragmented entries in refrain 5, and finally the slow and distorted forms in refrain 6—may be understood as tracking a process of waning resilience. The episodes bear out the reasons, in terms both of the dramatic events and their music.

The beginning of the first episode at [19 D-E], when the magistrate opens the trial and the chorus of onlookers confirms mother Kepler's arrival in the courtroom, is still laced with snippets from the subject. In the episode's longer second segment ([19 F-H]), the accompanying texture presents a musical portrayal of the magical thinking and rampant superstition exhibited not so much by the defendant herself but by prosecutor, witnesses, and crowd. The list of crimes these people are capable of imagining anybody like mother Kepler committing includes all the ingredients of a typical witch trial: she is accused of hexing, riding out of a chimney, twisting children's faces, bewitching a pregnant woman who then gives birth to a deformed baby, of causing the earth to split, causing a neighbor's cow's milk to go sour, etc.

How absurdly concocted these allegations are is expressed in the three musical strands that unfold simultaneously. Flute and three clarinets repeat a figure in four-part parallel; its chromatic ambivalence mirrors that in the refrain's fugal subject, while its $\frac{5}{8} + \frac{5}{8}$-length defies the reigning $\frac{3}{8}$ time and requires three repetitions to regain its initial metric position. Meanwhile, the lowest strings play a motif of giant leaps whose outlandish width (9th, 11th, and 14th) is bridged at the end by a chain of ascending fourths, while its metric irregularity is compensated in notable awkwardness by an accelerating rhythm. The resulting 8-bar phrase does not, however, line up with the woodwinds' 10-bar phrase. In a third strand, the spectators sing "Punish her!" in four transpositions of a bitonal scalar descent at decreasing distances. The popular verdict, sounding ever more urgent in the midst of the polymetric chaos, makes it clear that the trial is not aimed at ascertaining truth but at pronouncing a predetermined condemnation.

After the second refrain has established that all evidence is flawed and nothing can be proven, episode 2 starts over with new allegations. They are leveled by the four women who had seen mother Kepler by moonlight in the cemetery and now accuse her of whoring with the devil. Kepler's brother Christoph steps forward to say that while there had been no Satan—he himself had joined his mother that night to call her home—she was indeed guilty of desecrating a grave. Her counsel finds the involvement of a son reassuring enough to remind the court that the defendant is but a slightly confused old woman. In keeping with its less bizarre content, this episode is musically straightforward.

Yet once mother Kepler's own statement in the third refrain has refocused the attention on her herbalist skills, the defiant material of episode 1 recurs and she is shown the torture instruments. Her anguished singing of the Lord's Prayer in the fourth refrain is serendipitously answered by Kepler's entry into the courtroom. Upon his appearance on stage (at [19 U]), the lower strings launch a bass pedal on C♯ that then resounds through 64 measures, until mother Kepler is formally acquitted.

The central insert into the trial rondo (at [20]), which balances the moon rondo's hymn, features the exchange between Kepler and the magistrate, commented at each stage by the crowd. Extensive and musically self-contained, this segment is based on a collage of eight stacked strands (violas, violins, Kepler, high woodwinds, magistrate, brass, chorus, cellos + low woodwinds) that is repeated eight times with variations in the vocal parts.

The fifth refrain with its fragmented subject entries brings a jolt. Having gained his mother's acquittal, Kepler now faces reproaches from his brother, who would have liked to see her locked up safely and thereby prevented from any further mischief. Kepler is aghast at his brother's self-interest; Christoph becomes angry and leaves abruptly. After another transition (at [21]), in which mother Kepler deplores the curse of children who cannot accept that their mother simply thinks more clearly then they do, she embarks on what might have been the final episode but is in fact another wholly independent structure: a sonata rondo.

This second-order rondo is by no means small. Spanning 81 measures in a mood marked "Greatly moved and enthusiastic" plus a 13-bar codetta in slow tempo, it encompasses two episodes and two developmental sections:

refrain		refrain		refrain		refrain		refrain		refrain
	episode 1		episode 2		developm. section		episode 2a		developm. section	

The refrain is based on a lengthy thematic phrase that employs all 12 tones in a sequence deceptively laced with fourths and other highly melodic intervals. (For its instrumental exposition see [21 A]; the texted version in the original key appears at [21 F], in the fifth refrain.)

EXAMPLE 13: The refrain in Mother Kepler's "revelation rondo"

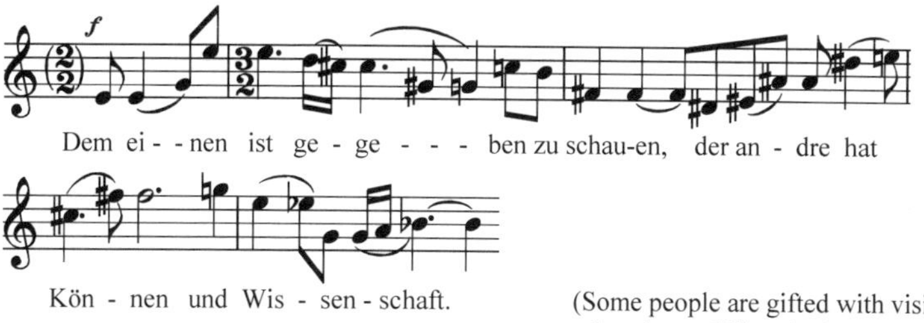

(Some people are gifted with visions, others have skill and science.)

All seven statements of the phrase either voice or accompany mother Kepler's revelation of her alleged power. Her most significant avowals are clad in a vocal version of the theme. (By contrast, her son's brief interjections are entirely restricted to the episodes.) As she tells him, he did not come to the trial and prevent her torture and death because of filial love and pity, as he claims, but because she conjured him with "a stronger power" ([21 B], theme on D). Now that the acquittal proves her soundness of mind, the moment has come to reveal her precious secret: she knows an herb that will cure all ills, and a metal more valuable than gold that will forever abolish the world's poverty ([21 D]). All she needs is a helper who can build laboratories for brewing, oversee the extraction of the ore, and distribute those elixirs of health and wealth. Thus what Kepler must do to partake of the bounty is simple: give up wife, children, and profession ([21 H], theme on D♯) and join her. She believes to be astute when she couches her suggestions in his own language and speaks of the "harmony" of his and her knowledge, which will "resound throughout the world." When her dismayed son refuses to hear any more, thus rejecting what she considers an offer of exceedingly lucrative collaboration, she is convinced that this could happen only because her younger son prevented her from digging up her dead father's skull in the cemetery and fashioning the goblet from which Kepler ought to have drunk.

The trial rondo's concluding refrain with its four entries of the slowed-down, rhythmically distorted fugal subject accompanies mother Kepler's dejected response. All is lost for her now that Kepler is not willing to assist her plan for the glorious improvement of the human (and especially her own) lot. Freedom means nothing to her; Kepler has achieved what the court could not: to drag her to the real scaffold. The trial rondo's coda, which concludes act III, recalls the incipit of mother Kepler's revelation theme. The curtain closes over three ***fff*** strokes of a pure B♭-minor triad—the tonal counterpole to the opera's tonic E, the fundamental of Kepler's search for harmony.

What does a hymn in the form of a fugue-passacaglia hybrid do in the middle of one rondo and a self-contained musical collage in the other? What symbolic messages are we to glean from a rondo whose refrains are designed as fragments of an ongoing fugue, and of a sonata rondo inserted as an "episode" into another rondo?

With their characteristic round-about motion, rondos allude fairly directly to the motions observed in the sky. According to a long tradition rooted in Plato's teachings, the circle was considered the most perfect geometric figure. Since God's Creation is by definition perfect, it was deemed proven *a priori* that figures like planetary orbits must also be circular. Copernicus and, even more so, Ptolemy required highly complex explanations to account for the less than straightforward courses of the planets as seen from the earth. It was Kepler who first considered that planetary orbits might be elliptical rather than circular. His conviction that the divine design must be simple and harmonious prompted him to speculate in ways that assumed basic ratios for every equation describing cosmologic data. This eventually led to his laws of planetary motion, which constitute a hymn to harmony—harmony in the sky, most literally, but also in all of God's Creation.

With the 8-part, 8fold musical collage that achieves the (dramatic) acquittal of a woman accused of witchcraft, Hindemith may be alluding to the 8 heavenly bodies known in Kepler's time, which are allegorically represented by 8 of the operatic soloists in the cosmological transfiguration that crowns the opera. One among these eight differs from the others in important respects: the moon. Its orbit alone is indeed circular. Furthermore, as Hindemith reminds us with his rondo inside another rondo, it is the only cosmic body that travels in a loop within another loop: circling around the earth while accompanying it on its course around the sun.

Mother Kepler's surreal notions stem from the pen of the imaginative librettist; Kepler's biographers do not seem to have heard of such an outcome of the witch trial. But as the music shows us, these notions constitute something akin to "a view of earthly reality from the lunar perspective" —and thus an ingenious counterpart to her son's *Somnium*.

Hesse's Castalian "Chinese House Game"

It is not immediately obvious that geometry, or the shapes, measurements, and relationships of figures, might serve as a key to a literary text—especially one that aims at depicting a world built around the self-imposed task of preserving the treasure trove of all significant cultures through a highly sophisticated play with symbols and ciphers. And yet this science, which ancient thinkers knew to be crucial for an understanding of the created world in its material as well as spiritual dimensions, plays a significant role at a pivotal point in the Castalian story.

As many Hesse scholars and readers have observed, the glass bead game is tantalizingly elusive. At the same time, it is allegedly all-important to the self-definition of the World of the Mind and representative of its aims. The latter function is fulfilled most notably with the annual ceremonial game, the *ludus solemnis*, which draws thousands of fascinated spectators ("passive participants") from non-Castalian walks of life. According to the fictional chronicler, one of the most memorable official performances in history was the first *ludus solemnis* Knecht composed and celebrated after his election as *magister*. But once again, readers learn more about the impression created and the praise received than about this game's actual content and form. The narrator offers merely a brief remark, which seems intended to tease or exasperate by withholding any concrete information: that Knecht's "pretty idea" had been "to base the structure and dimensions of the Game on the ancient Confucian pattern for the ritual building of a Chinese house" (p. 245). It is left to readers to follow the hint, learn more about these transcultural patterns, and seek to understand how they might inform a game.

Much more than human homes in other cultures, traditional Chinese dwelling houses were built not only with practical, architectonic, and aesthetic concerns in mind, but in an elaborate pursuit of harmony: the external harmony with the cosmos and the spirit of the place as well as the internal harmony in the form of spatial proportion and respect for religion and social order. Knecht is said to have developed the game plan in connection with a thought he had during his study of the *I Ching*: that "the mythic order and significance of the rules made an unusually appealing and charming symbol of the cosmos and man's place in the universe" and "the age-old mythic spirit of the people [...] seemed to him wonderfully and intimately fused with the mandarin and magisterial spirit of speculative scholarliness" (p. 245). Figures and spaces, along with their material and spiritual size, proportion, and place, may open an unexpected door to the game not so much in terms of its technique or procedure, but with regard to its ideal content and dialectic.

In ancient Chinese belief, earth, water and wind correspond to the human soul. They do not merely mirror it but *are* the soul itself—or, phrased differently: the soul is part of them and contained in them. It follows that the two most important buildings in human life, the family home and the tomb, were to be designed in accordance with the spiritual forces of surrounding hills and bodies of water. Building, especially digging, had to be undertaken with explicit respect for the earth, whose body should never be injured and scarred without due apologies and signs of gratitude. The good fortune a house would bring its inhabitants depended on the observance of astrologically determined days for the beginning of the various stages in the complex construction process, as well as on the layout of the house itself.

The best source of information on this subject is the *Lu Ban jing*, the "Classic of Lu Ban," a carpenter's manual compiled in the 15th century in honor of Lu Ban, a master craftsman from the time of Confucius who later became a kind of patron saint for carpenters and builders.[11] Although perhaps not on a par with the great texts of the Chinese cultural tradition, its editors have accorded the book the rank of a "Classic": the title shares the generic component with works such as the *Dao De jing*, the *Yi jing* (better known in its old transliteration as *I Ching*), the *Shi jing* (Classic of Poetry), etc.

The *Lu Ban jing* is an extensive manual. The English version contains 150 pages of translation complemented by the translator's introduction and supporting material. Of its three sections, the first addresses myriad aspects relating to the building of vernacular dwellings (50 pages, including a brief extension on palaces, temples, shrines, and garden kiosks), the second focuses on the making of furniture and implements (73 pages, including a 15-page opening section on structures in yard and garden), while the 23-page third section informs about "Secret Charms and Magical Devices." The apparent separation of craft and spell is deceptive though: as the reader soon discovers, the construction of a dwelling house was a complicated process in which ritual and technique were so closely linked as to seem virtually inseparable.

Moreover—and this is why the topic may have offered itself to Knecht /Hesse—the building process of a vernacular home involved all areas of harmony: music,[12] arithmetic, geometry, astronomy, astrology, religion, and those visual arts that are defined by a concern with proportion. The metaphysical background, captured in the *I Ching*, deserves a separate discussion

[11] Klaas Ruitenbeek, *Carpentry and Building in Late Imperial China: A Study of the Fifteenth-Century Carpenter's Manual* Lu Ban jing (Leiden: Brill, 1993).

[12] Especially impressive are the "pounding songs," ballads with up to 600 verses that accompanied the preparation of the foundation; see Ruitenbeek, *Carpentry and Building*, p. 49.

(see part V below) as it informs not only this particular game but also the larger theme and goal in Hesse's work. Among the other features Knecht is reported to have found so attractive, several relate to the meaning attributed to appropriate spacing, placing, and timing—the "orientation by the points of the compass, the gates, the spirit wall" (p. 245)—and to rituals. Whereas Western culture might expect to encounter these especially in connection with the building of temples and churches, Chinese culture reserved the most elaborate considerations for private dwelling places. As Ruitenbeek remarks: "While it is true that certain rituals and ceremonies did accompany imperial and governmental building, it is only in the case of vernacular architecture that almost every step in the building process called for ritual acts."[13]

No respectful Chinese ever undertook any construction without seeking the advice of a geomancer, who determined the adequate site for the house or at the very least its direction in relation to the cosmos. All parties—owner, geomancer, and carpenter—also consulted the almanac. Chinese almanacs were similar to, yet in one respect distinctly different from, the ones Kepler and his fellow astronomers were required to compile each year in 17th-century Europe. Both kinds were dominated by entries related to farming. But where the secondary focus in Western almanacs was directed to the more or less favorable days for military and diplomatic ventures, their Chinese counterparts revolved around house building. Entries referring to construction outnumbered those in all other non-agricultural categories taken together. This emphasis reflects the high degree of ritualization so characteristic of house building in China. The *Lu Ban jing* stipulates that timing in accordance with the current almanac must be observed for twenty of the steps in the construction of the house itself,[14] for five kinds of additional structures in garden or yard, and when "placing the bed and fitting the curtain."

The spatial aspects of geomancy consider features in the landscape—bodies of water, mountains, graveyards, etc. (this part, along with its watered-down cousin currently en vogue in the West, is known as *feng shui*)—as well as the relationship of the building's ground plan to the "stars." The distribution of these "stars"—which according to Ruitenbeek are fictional, not to be found in any sky—is different each year, and announced in the almanac. Unfavorable directions must absolutely be avoided for the two

[13]Ruitenbeek, *Carpentry and Building ...*, p. 5.

[14]These range from "going to the mountains to fell trees" and "cutting bamboo without the risk of it becoming worm-eaten" through "piling up the wood and making a shed," "stirring the ground to start building," "marking the columns," etc. to "plastering walls," "paving the floor," and "making stone stairs" (Ruitenbeek, *Carpentry and Building ...*, pp. 43-44).

endpoints of the home's longitudinal axis, i.e., the positions of its rearmost part, the "seat," and its front gate, the "orientation" or "view." Moreover, whenever any finishing work or repair was undertaken after the home owner's ancestral tablets had already been moved into the house, care had to be taken that the work did not take place in a direction from the ancestral altar that the current almanac declared unfavorable. Should intended building work involve an inauspicious direction, one could either wait for a year in which this direction was favorable, or temporarily transfer the center of the house to a position that allowed a favorable reading of the geomancer's compass (see fig. 21). Similarly, when the timber arrived at the building site, it had to be piled up in a favorable direction, and ritual specifications determined the handling and movements of the ridge pole before and while it was hoisted onto the roof.

FIGURE 22: Chinese compass card

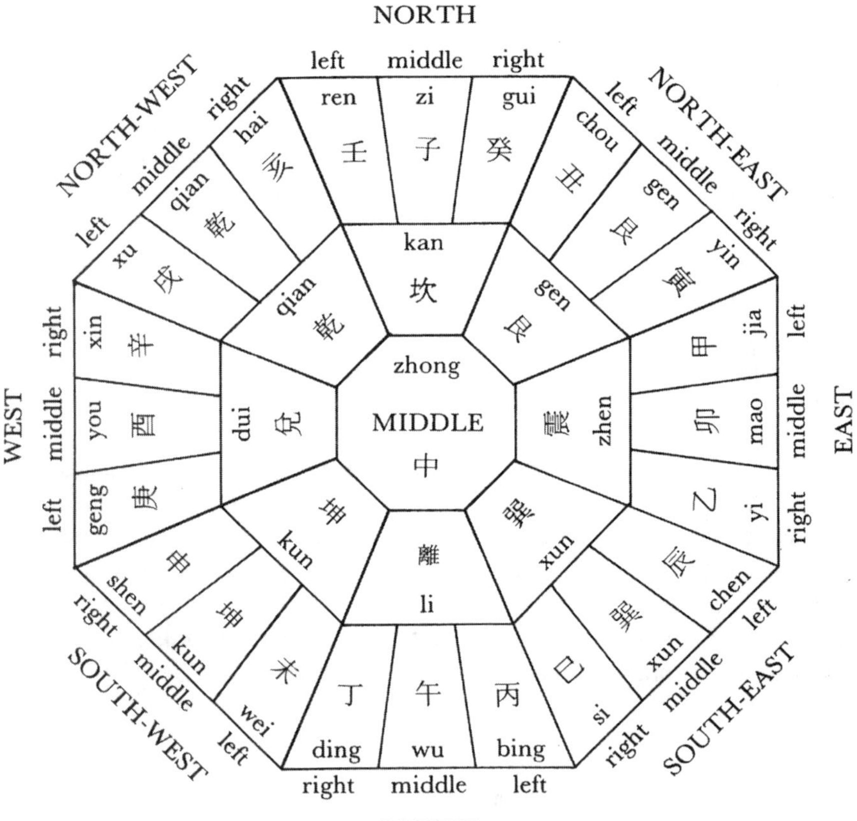

(from Klaas Ruitenbeek, *Carpentry and Building in Late Imperial China*, p. 56).

The designated center of any dwelling house was the altar table, which normally held the ancestral tablets together with offerings of incense, candles, and flowers. During times of construction, these items were replaced by (or, during repair work, complemented with) a portrait of the patron saint Pu An, offerings of food and wine, and an assortment of carpentry implements. Although major parts of any house were made from earth, wood was considered its "real" building material. One reason for this self-deception was the perceived connection between, and opposition of, *yang*- and *yin*-houses: *yang*-houses were the dwellings for those living above ground, to be thought of as made of wood, while *yin*-houses sheltered those living underground (the dead) and were therefore to be made of earth and stone. Geomancy and popular belief saw a close link between tombs and houses: the ancestors' bones were kept in tombs while their tablets—and thus their spirits—resided inside the home. Similar reasons were given to justify attention on favorable timing and measurement for a single item of furniture, the bed: "it is the place where those who, in the future, will minister to the ancestral altar of the house and, in the long run, even the ancestors themselves are created."[15]

Measurements were subject to ritual as well as aesthetic considerations. Ruitenbeek reports that ideally, the number of tools carpenters were to bring to a house-building site should be twenty-eight, corresponding to the constellations in the sky. Doors especially must correspond to one of several favorable measurements. Yet these were theoretical absolutes: a choice of carpenter's rulers with varying unit widths allowed for all kinds of desired door sizes to correspond with favorable measurements according to one scale or another. Other measures reflected proportions and resembled Western ideas of numerical harmony. The slope of a roof followed rules such as: "for each foot of depth, the roof is lowered by five [Chinese] inches," resulting in a ratio of 1:2.

With regard to floor plans, the overriding consideration was symmetry and proportion. As confirmed by Rudolf Kelling, a 20th-century architect measuring the oldest vernacular buildings in rural China, even the smallest peasant homes, consisting of nothing but two rooms, were laid out in accordance with simple ratios. The main room would typically be built over a square, the bedroom would be of equal depth but half the width, so that the perimeter of the entire building corresponded to the ratio 2:3. If a verandah was added to the front, its depth would relate to that of the room behind it as 1:4 or 1:3, etc.[16]

[15]Ruitenbeek, *Carpentry and Building ...*, p. 89.

[16]On this see the various floor plans in Rudolf Kelling, *Das chinesische Wohnhaus* (Tokyo: Deutsche Gesellschaft für Natur- und Völkerkunde Ostasiens, 1935), pp. 9-15.

FIGURE 23: Design of a Chinese courtyard house (bird's eye view)

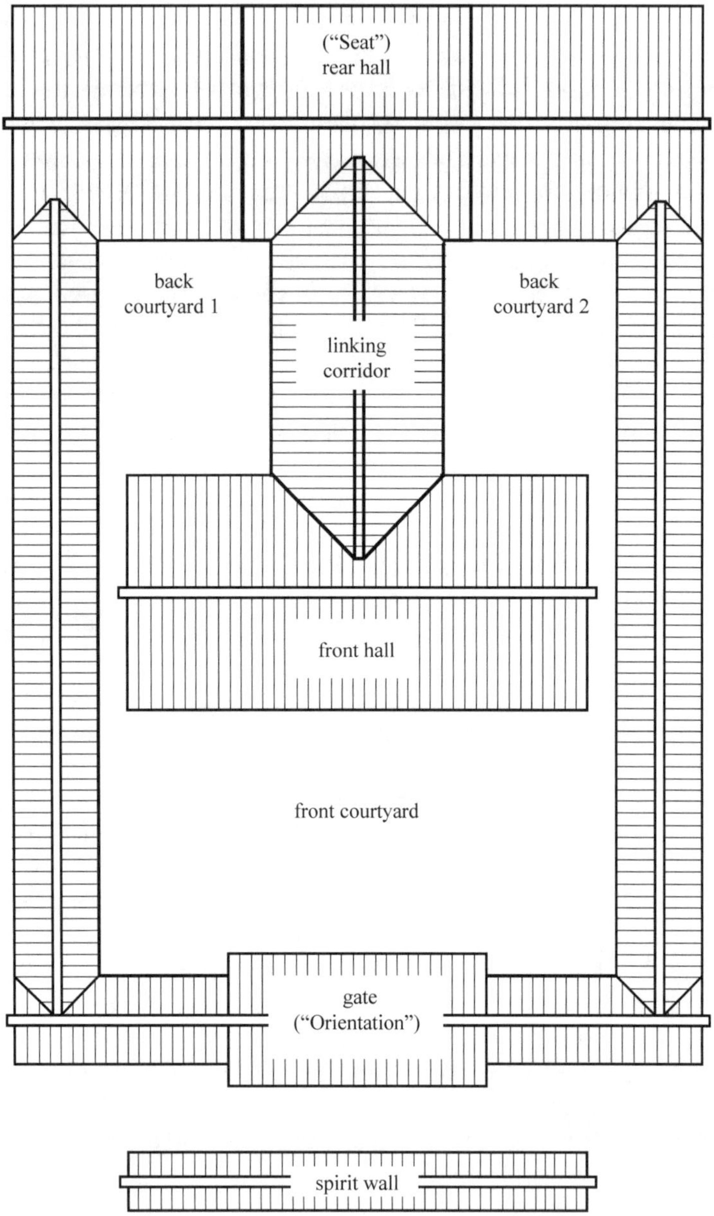

(after Klaas Ruitenbeek, *Carpentry and Building in Late Imperial China*, p. 60; spirit wall added)

Ruitenbeek's *Carpentry and Building in Late Imperial China* includes a ground plan of a typical dwelling house (see figure 22). It shows the characteristic H-shape for the main quarters, three courtyards, and an enclosing wall with ambulatory and gate. It is perfectly symmetrical and—he does not tell but I measured and confirmed my hunch—all proportions are harmonious in the Chinese *and* the Pythagorean sense:

- The rear hall and its two adjoining rooms are squares;
  also, the rear hall, the linking corridor, and the
  front hall are identical in depth (ratio of 1:1);
- 1:2 defines the shapes of the front hall
  and of the front courtyard;
- 1:3 is the rear hall's relation to the compound's width;
- 2:3 pertains to the perimeter of the whole compound
  as well as to its gate,
- 3:4 to each of the two rear courtyards
  as well as to the linking corridor,
- 3:8 to the relative widths of corridor : front hall
  and gate : total compound.
- Finally, the width of the horseshoe-shaped ambulatory is
  1:8 / 1:12 of the compound's width / depth.

The seemingly unlikely combination of lucky numbers and auspicious dates with pounding songs and furniture ornaments on the one hand, proportional spacing, geologic adaptation, and cosmological alignment on the other, expresses a concern for all-encompassing harmony. A glass bead game based on these principles and developing, say, two contrasting themes and three supporting motifs, could certainly be designed. Based on the recurring justifications given for the various rules in the *Carpenter's Manual*, one might imagine it as turning, for instance, on the motifs of body, mind, and spirit: The *body* (of each individual inhabitant and of the family as a whole) is carefully placed in relationship to features of the earth's body. The family home's only external "orifice," the gate, is protected against evil winds or other forces with a spirit wall. The *mind* is pleased and uplifted by the sense of order expressed in the perfectly symmetrical and proportional design. The inner consonance manifested in this interrelatedness of differently sized entities mirrors the design of the social realm, where the individual is ideally a harmonious part of a family, the family, a harmonious part of a community, and the community, a harmonious part of the nation. The *spirit* of the individual, but even more of the family as a whole, is served at the ancestral altar—typically placed in the rear hall, the "seat," which thereby becomes the home's spiritual center—and through respect for the "stars" and their impact on human decisions.

The two themes on which a glass bead game "based on the ancient ritual Confucian pattern for the building of a Chinese house" might be founded—and which, in true Castalian spirit, ought to be developed toward synthesis—can be imagined as something like "submission" and "art": submission as the well-reared individual's readiness to serve as an integral part in a hierarchical order, and art as a treasure trove of cultural heritage. These happen to be, it will be noticed, the implicit themes of the Castalian "game" as a whole. Early in the introduction, the chronicler informs readers about the former: "For us, a man is a hero and deserves special interest only if his nature and his education have rendered him able to let his individuality be almost perfectly absorbed in its hierarchic function without at the same time forfeiting the vigorous, fresh, admirable impetus which make for the savor and worth of the individual" (p. 13). A little later, he explains the latter, i.e., the Castalian attitude toward the arts: "Ever since we have renounced [...] trying to vie creatively with those generations, ever since we have also forsworn [...] that purely sensuous cult [...], ever since then have we been able to understand, more purely and more correctly, the general image of that culture whose heirs we are" (pp. 27-28).

Several aspects of traditional Chinese house building relate to what I suggest as the game's principal theme: the individual's submission to the good of a larger entity. The formal layout of the building with its spatial hierarchy reflects both the Confucian teaching of social hierarchy within the family and the temporal hierarchy of the ancestral line. A thoughtful placement of the house in relationship to the features of the earth expresses the Daoist belief that humans ought to live in close harmony with their natural surroundings. The eagerness to orient both the timing of all construction work and the home's ultimate direction to the "stars" testifies to the belief that the human world is merely a microcosm within a much larger cosmic order.

The secondary theme, art, was equally central to the building of a traditional Chinese house. Strictly speaking, this was art appreciated not for the sake of sensual or aesthetic pleasure but as a system of cultural and edifying symbols; but the ideal attitude did not have to exclude the affective appreciation. Music, poetry, and the visual arts all had their specific time and place in a Chinese dwelling house.

Music occurs in connection with all work requiring temporal alignment of several workers, but its benefit never stops there. The example explicitly mentioned is the "pounding song." On the surface, it served a straightforward purpose: the length of the ballads sung by the earth workers during the preparation of a new building's mud floor and walls determined the time spent with the task and thereby, the solidity of the foundation. Interestingly though, these pounders were not singing raucous popular songs. The source Ruitenbeek cites is a ballad from the Ming Dynasty, describing the quest of a heroic ancestor of masons and pounders for his father. In other words: the singing connected the builders and, through them, the family and the larger community to their historical or mythical past.[17]

Visual art of three kinds prevailed in the Chinese house: furniture and window or door panes would typically contain carvings; framed water colors or calligraphy might decorate walls; and flowers, shrubs, or trees growing in the courtyards or between a window and the outer compound wall were often artfully pruned so as to create seemingly "natural" pictures when seen through an opening. In all cases, the immediate purpose might seem to be aesthetic, but the ultimate end was moral. Furniture carving, the item most closely related to house building, comes with three themes: abstract arabesques or vignettes, stylized arrangements of flowers and birds or stylized dragons, and scenes from Chinese mythology. The three types were intended to remind the house's inhabitants, respectively, of the world's comforting if often labyrinthine order, of the human interconnectedness with visible and invisible nature, and of each individual's embeddedness in an old culture whose moral beliefs are captured in edifying stories. What appears as embellishment thus serves in furthering the psychological foundation of the house: reliance on order, serenity, and respect for the past and its lessons.

Poetry was recited at various stages of the carpenters' work, notably but not only when the ridge pole was placed; it also accompanied the inhabitants' daily lives in the form of couplets pasted next to the vertical beams of doorways, and was often a part of the art work decorating the walls of family rooms. All three kinds of poetry had the spiritual aim of enhancing mindfulness. Furthermore, they made the writings and philosophy of old masters relevant to contemporary life.

Finally, the walled-in garden spaces were another contribution to the theme of art. Classical private gardens in China were conceived as cultural rather than scenic. Often entrusted to artists and poets, they were regarded as three-dimensional landscape paintings, or landscape poetry materialized. The idea was to provide a meditative space set apart from the everyday reality of social life. Many of these gardens in fact resembled the paintings on the walls inside the house, in that they artfully recreated the poetic genre of landscape pictures and aimed at a universal rather than a sentimental and specific impression.

The two themes of my speculative *ludus* "based on the ancient Confucian pattern for the ritual building of a Chinese house"—the individual's

[17]Ruitenbeek (p. 49, note 135) mentions several manuscript copies with variants of the ballad.

submission to a hierarchical order and the arts' contribution to people's physical, emotional, and spiritual grounding—would thus have allowed Knecht to find Castalian philosophy mirrored in an ancient Chinese tradition. While the themes are not strictly in contrast to one another, the rules of the glass bead game would have required that they be developed toward a kind of synthesis, a larger harmony and unity.

This synthesis is achieved in *ritual*. For that is what auspicious dates, lucky measures, and favorable compass fields are really about. They serve as a reminder that no human is the master of his or her own fate, at the same time giving expression to the belief that a person can, through an attitude of respect and a variety of formal placating acts, contribute to a best-possible outcome of various undertakings. Similarly in Castalia, ritual, the final step in the developmental process of the glass bead game, has become what to outsiders might appear as its most striking feature. When participants of the annual ceremonial game obey strictly regulated exercises of fasting and meditation to prepare themselves for the exalting experience, when they let themselves be entranced by the ritual gestures with which the *magister ludi* conducts the steps of the game, they sense that the outcome for every individual is more than the sum of the intellectual edification and the insight induced in guided contemplation. Here, too, human beings are mysteriously connected with the larger Oneness.

Joseph Knecht's first ceremonial game, "remembered to this day as the Chinese House Game and often quoted" (p. 262), must be understood as an expression of his attitude toward Castalia at the height of his career. Players and guests alike would have relished finding the principal tenets of the Pedagogical Province endorsed and corroborated in the age-old tradition of the Chinese classics.

Part IV
ASTRONOMY

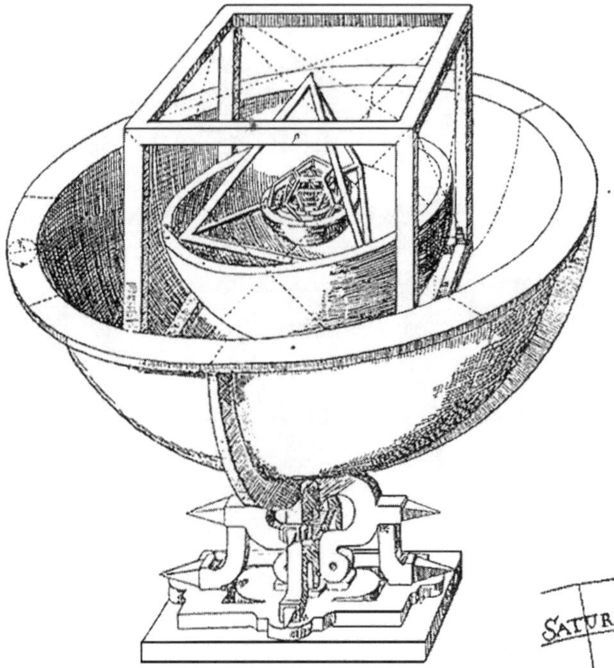

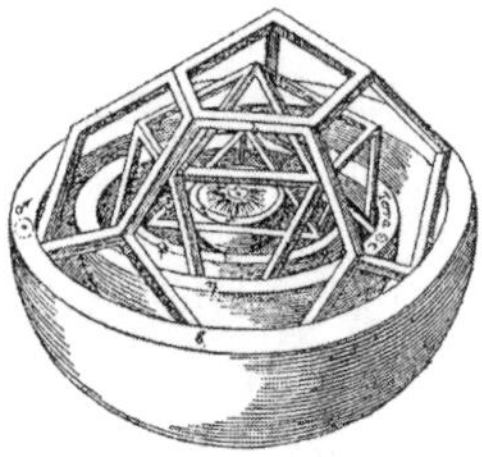

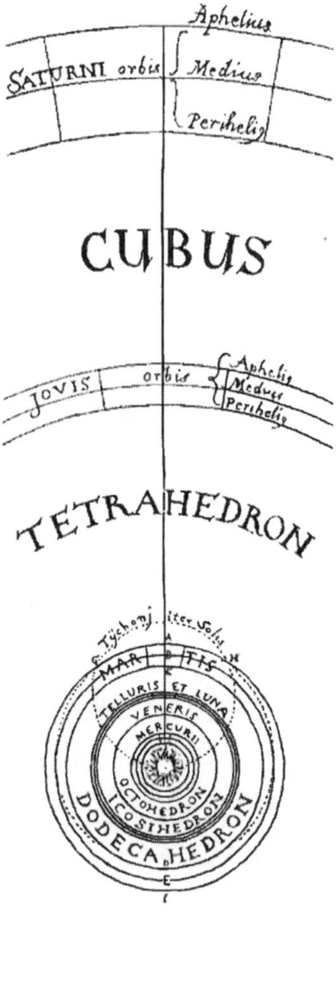

PLATE 4:
Kepler's representations of interplanetary distances as defined by the Platonic Solids

The Divine Signature in Cosmic Harmony

The Pythagorean philosophers perceived the universe as created on the basis of features that would grant *harmony*. According to them,

- the cosmos is alive and endowed with a soul; it is an ordered entity, governed by mathematical laws, in which each component has a specific function;
- the earth is spherical, hovering motionless in the center of the cosmos;
- the sky turns clockwise around an axis that runs through the earth;
- the sun, the moon, and the planets, besides being carried along with the motion of the sky, also turn independently, counterclockwise in circles whose planes are tilted against that of the astral motion; they move at various distances from the earth, their sequence from the center outward being: moon, Venus, Mercury, sun, Mars, Jupiter, and Saturn,[1]
- the speeds at which each of these luminous bodies travels along its orbit relate to one another in whole numbers;
- after a number of years corresponding to the smallest common multiple of all orbital speeds, the sun, the moon, and the planets will recur in the same position with regard to one another and to the stars;[2]
- each of the planets emits in its motion a tone, which together with the other tones creates a harmony.

Philolaus, an astronomer who flourished in the 5th century B.C.E. and was a follower of Pythagoras, modified this theory somewhat. He postulated that earth, sun, moon, and planets all moved around a central fire hidden from view by an interposed counter-earth. Another Greek thinker, Aristarchos of Samos, believed that motions in the sky could be explained by the hypothesis that the earth describes a full turn on its axis in the course of every 24 hours and that, along with the other planets, it moves around the sun. This explanation was, however, rejected by most Greek philosophers,

[1]Plato, presenting his astronomy in two corresponding passages of his *Republic* and *Timaeus*, differs in this point; he places the sun nearer to the earth, between the moon and Venus.

[2]This "great year" or "Platonic year" is assumed to last for ca. 25,920 years; a "great month" then takes ca. 2160 years, calculated in disregard of the different sizes of the zodiac signs to which these "months" are linked due to the slowly preceding vernal equinox point.

who regarded the large, heavy earth as a motionless globe around which the light, incorporeal celestial bodies revolve. In the 2nd century C.E., Claudius Ptolemy of Alexandria proposed a model of the universe in which each celestial body traveled on a small circle called an epicycle, centered on a point that moved around the earth in a large circle called a deferent. The model depicted the motions of the heavenly bodies fairly accurately, but offered no physical explanation for them. Nonetheless, Ptolemy's model remained in place for almost 1500 years.

The basic assumptions about the earth's centrality and immobility were first questioned by the German philosopher and mathematician Nicholas of Cusa and the Italian artist and scientist Leonardo da Vinci. In the 16th century, Nicolaus Copernicus developed the heliocentric model of the solar system, in which the sun is stationary at (or rather, very close to) the center. The planets, among which the earth is now reckoned, travel around it. This view of the solar system challenged the earlier geocentric model, in which sun, moon, and planets are all quite different in nature from mother earth, around which they move. Though it was almost as complex as the Ptolemaic system, the Copernican model gradually gained acceptance. It achieved final victory in the early 17th century, owing to discoveries made by Tycho Brahe and Galileo Galilei with the new astronomical telescope and the development of a new physics.

Neither Ptolemy nor Copernicus make any reference to the physical causes *why* planets should move the way they do. In the absence of any such explanation, it was considered the nature of a celestial body to move in a circle (this being the most perfect geometric figure), just as Aristotle had declared it the nature of terrestrial objects (such as falling stones) to move straight downwards, towards their "natural place" in the earth's center. By the time Kepler entered the field, the preconditions had changed. Tycho Brahe had collected astronomic data of unprecedented precision, and William Gilbert had proven the earth's magnetic efflux and analyzed its power to attract and repel other magnetic bodies. Kepler, explaining his fundamental point of view in *The Epitome*, gives credit to the Polish, Danish, and English ancestry of his insights when he states: "I build my whole astronomy upon Copernicus' hypotheses concerning the world, upon Tycho Brahe's observations, and lastly upon the Englishman William Gilbert's philosophy of magnetism."[3] Brahe's data for Mars had forced Kepler to acknowledge that the movements of this planet at least could not be

[3]Kepler, *Epitome of Copernican Astronomy, books IV and V, and The Harmonies of the world: V*, transl. by Charles Glenn Wallis (Chicago: Encyclopaedia Britannica Inc., 1952), p. 850; W. Gilbert, *Of Magnets, Magnetic Bodies, and the Great Magnet of Earth* (1600).

explained on the basis of circular orbits; only elliptical trajectories, whereby the sun was understood to occupy one of the ellipse's two foci, yielded a plausible explanation. The fact that Kepler sought an explanation, rather than stating mere observations, documents a quantum shift in the approach to the field of astronomy: henceforth, for an observation to be of interest, it had to be in conformity with physical laws that would elucidate its causes.

Kepler's new approach took care of several puzzling aspects: The planets' paths, their varying distance from the sun, and their varying speed. Kepler explained all these features as the result of a physical mechanism by which the sun's magnetic force sweeps the planets around their orbits.

At the time the 25-year-old Kepler published his first work on astronomy, *Mysterium cosmographicum* (the full title translates as Precursor of Cosmographic Dissertations, or The Cosmographic Mystery), he did not yet have Brahe's very accurate data and had no reason to doubt the age-old conviction that the planets travel in circular orbits—albeit no longer around the earth but, since Copernicus, around the sun. What he set out to do as early as 1596 appears in retrospect as a first attempt to reveal a geometric harmony in the structure built by the heavenly bodies then known. At this point, he did not address angles or orbital speeds but relative position and number. Why was the earth, traveling around the sun, joined by just five other planets? And why are their distances from one another just the way they are? Kepler believed that there must be a rationale and sought it in geometry. Geometry knows precisely five regular bodies, the solids: three-dimensional figures composed of identically shaped faces that join at identical angles. He fitted the solids into the five intervals between the six orbits in an order determined by their importance. Since polyhedra are closer to the spherical shape the larger the number of their surfaces, whereas a tetrahedron or cube forces a sphere placed around it much further apart from one fitted inside it, the distances created by the interpolated solids are now understood as resulting from geometric laws. Kepler's chief discovery in *Cosmographic Mystery* consisted in determining a sequence of planets alternating with solids in which the relative sizes of the orbits would correspond with the radii of the planetary spheres given by Copernicus. The order, he stated, was:

Saturn–cube–*Jupiter*–tetrahedron–*Mars*–dodecahedron–*earth*–icosahedron–*Venus*–octahedron–*Mercury*.

The fit was close enough to be suggestive and to recommend the young professor of mathematics to the professional world; it was imperfect enough not to let him rest but inspire him to continue his explorations.

In 1600, Kepler went to Prague to work with Tycho Brahe; a year later, Brahe died and Kepler followed him as imperial mathematician. Analyzing

Brahe's data for the motion of Mars, he convinced himself that this planet did not travel on a circular but on an elliptical orbit. He applied the same hypothesis to the other planets and discovered that their paths, for which astronomers could account only with composite explanations, seemed far more logical under this new assumption. This dethroned the circle, the most perfect geometric figure.

Another speculation that was also counterintuitive to a thinking that presupposed a Divine Geometer—that the point around which planets orbit be "central"—proved particularly fruitful. Kepler assumed that the sun is not positioned on the cut of the ellipse's two axes but in one of its two foci, thus exerting different powers of magnetic attraction at different times because of the vastly differing distances. This hypothesis solved the remaining questions regarding path and speed. In his next major treatise on astronomy, *The New Astronomy* (1609), Kepler published what became known as his first and second laws of planetary motion: (1) planets move in elliptical orbits, with the sun at one focus; (2) the radius vector from the sun to the planet sweeps over equal areas of the ellipse in equal times.

The decisive third law of planetary motion did not emerge until shortly before the conclusion of Kepler's *Harmonices mundi libri V*, completed in 1618 and published a year later. In this work, Kepler had returned to his aim of a quarter-century earlier: to prove that the universe was ordered in such a way that the all data would be perfectly "harmonized." After elaborating in much detail how the same kind of harmony is at work in the other aspects of life on earth—in the abstraction of arithmetic and geometry just as in the concrete world of music and architecture—he argues that it must also apply to heavenly motions and positions. And if numbers and the shapes and events they define can be imagined as embodied in musical intervals, each planet then had to have its own range of notes. Orbits with a great degree of eccentricity, like that of Mercury, would necessarily "sing" more notes than those whose paths are almost circular. (The earth, as was mentioned earlier, is restricted to the two notes *mi* and *fa*—which Kepler deemed appropriate for humankind's home planet where, as everybody knew, *MI*sery and *FA*mine reign.)

In order to find a useful index of planetary motion in which to express the innate harmonies, Kepler drew up and tried out five hypotheses. He concluded that the correct factor was each planet's speed at the extreme points of their orbits around the sun. Harmonic relationships resulted from the ratio of the extreme speeds of neighboring planets. Since a planet's speed at any time is a function of its distance from the sun at that moment, it travels most slowly at its maximum distance from the sun (the so-called *aphelion*) and is fastest at the other endpoint of the major axis (the *perihelion*). Kepler measured those speeds by the arc the planet traverses at perihelion and aphelion, as seen from the sun, during 24 hours of earth time. He then calculated two ratios for each pair of planets: the interval between the perihelion (maximum) speed of the outer and the aphelion (minimum) speed of the inner planet, and that between the aphelion speed of the outer and the perihelion speed of the inner planet.[4] He found these ratios to correspond almost perfectly with those that define the basic musical intervals.[5] Here is Kepler's ten-part table; I have anglicized the speed-point specifiers and complemented the ratios with translations into the corresponding musical intervals:

Saturn fastest–Jupiter slowest	1 : 2	octave
Saturn slowest–Jupiter fastest	1 : 3	fifth + 1 octave
Jupiter fastest–Mars slowest	5 : 24	minor third + 2 octaves
Jupiter slowest–Mars fastest	1 : 8	3 octaves
Mars fastest–earth slowest	2 : 3	fifth
Mars slowest–earth fastest	5 : 12	minor third + 1 octave
earth fastest–Venus slowest	5 : 8	minor sixth
earth slowest–Venus fastest	3 : 5	major sixth
Venus fastest–Mercury slowest	3 : 5	major sixth
Venus slowest–Mercury fastest	1 : 4	2 octaves

Considering that Kepler's measurements rely on 17th-century instruments, they hold amazingly well over time. Various 20th-century physicists and astronomers have undertaken to check his ratios. The French scholar Francis Warrain was the first to focus on the planets' angular velocities.[6] He was able not only to confirm Kepler's harmonic interpretation, but to expand it to the three very remote planets that were not yet known in Kepler's lifetime, stating that the maximum and minimum speeds of Uranus, Neptune, and Pluto also relate to on another at harmonic ratios. Compare, e.g.

[4]The speeds Kepler measured are: Saturn: aphelion 1'30", perihelion 2'15"; Jupiter: aphelion 4'30", perihelion 5'30"; Mars: aphelion 26'14", perihelion 38' 1"; earth: aphelion 57' 3", perihelion 61'18"; Venus: aphelion 94'50", perihelion 97'37"; Mercury: aphelion 147' 1", perihelion 384' 0".

[5]Kepler admitted one exception to this perfect correspondence: the converging interval between Jupiter and Mars deviates from a musical consonance by almost a quarter-tone. This, however, is a discrepancy that is smaller than the degree by which the human voice in figured melody is almost perpetually out of tune, and by which tuned instruments go out of tune in the course of the performance period before they are readjusted.

[6]Francis Warrain, *Essai sur l'Harmonices mundi ou Musique du monde de Johannes Kepler* (Paris: Hermann, 1942), vol. 2, p. 136 and *passim.*

Uranus slowest–Saturn fastest	3 : 5	major sixth
Uranus fastest–Saturn slowest	5 : 6	minor third[7]

Kepler revered Plato; there are numerous references in his *Harmonices mundi* to the Platonic concept of "idea" or "original image." To discover appropriate proportions in the world of sense objects implied, to his mind, to reveal the analogy of these external proportions with a certain original "image" existing in the human soul, the "idea" of true and genuine harmony, which the sense objects only helped to bring to light. The proportions obtaining in the observable world were thus indicators for dispositions reigning in the soul.[8]

A terminological distinction may be in order here: what today's scientists measure may lead to the formulation of laws; what Kepler stated when he spoke about the proportions pertaining to celestial arrangements and motions is better understood in analogy to "norms." In his teleological perspective, Kepler expected his findings to yield a kind of morphological proof for God's existence. The prevalence of harmonious proportions in all aspects of created nature served him as a confirmation that all was the work of a Creator God; hence his feeling that artefacts created by humans in analogy, i.e., observing the same proportions, approach the divine. As Kepler understood the norms he discovered, they demonstrate not only the Creator's skill but, more importantly, His *intent*: harmony is not the result but the reason for the nature and astonishing arrangement of the planetary system—its *final cause*.

[7]A further expansion of the theory of "celestial harmony" should be mentioned here. Jean-Pierre Nicola's *Nombres et formes du cosmos* (Paris: Ed. Traditionnelles, 1973) and *Éléments de cosmogonie astrologique* (St-Denis-sur-Huisne: C.O.M.A.C., 1992) take their inspiration from Warrain. He uses diverse numerical transformations of astrometric data, including a planet's linear speed, average distance from the sun, average surface gravity at the equator, etc. On the basis of these calculations he claims to reveal a harmony not only between adjacent orbits, but in the larger group. Taking moon-Mars as an axis of symmetry, he proposes a grouping of planets into the pairs sun-Pluto, Venus-Neptune, Mercury-Uranus and Jupiter-Saturn (*Éléments*, p. 140, fig. 18). Other evidence suggests a possible alternative division into the three groups sun-Venus-Mercury, Jupiter-Mars-Saturn, and Uranus-Neptune-Pluto (*Éléments*, p. 145.)

[8]Interestingly, Kepler uses not only Plato's term "idea" but also, especially in the last-mentioned context, the word "archetype" which, 300 years later, Carl Gustav Jung would employ for the images shared by all humans in a collective unconscious, manifested in dreams and myths.

Cosmic Nesting and Orbiting in *Die Harmonie der Welt*

In his portrayal of Kepler and his search for universal consonance, Hindemith assigned celestial bodies—sun, moon, and the six planets then known—to the main characters. The protagonist himself is Earth. Like everybody else and from necessity rather than by choice, he orbits about the center of power, Sun, impersonated by the successive emperors, Rudolf and Ferdinand (who are played by the same singer). Kepler's wife Susanna is Venus, the embodiment of love; his superstitious mother is Moon. Ulrich, Kepler's indolent research assistant whom stubbornness and disappointment later turn into an unhappy soldier, is Mars; the two pastors who, at Linz in 1613 and at Regensburg in 1630, attempt to threaten Kepler into changing his unorthodox views about the nature of Christ's presence in the Eucharist, are allegorized as Mercury. As the innermost planet, Mercury may be understood either as a symbol of all that derives its strength from the second aspect of centralized power at the time, the Church, or as one who touched the core of Kepler's being. Wallenstein, the ambitious and high-handed general, is Jupiter, the planet named after the supreme god in the Roman pantheon. Finally there is Tansur, a Wittenberg University drop-out who survives on half-knowledge and recklessness, begins as a pitiful dabbler in Kepler's profession when he makes a living selling astrological prognoses and rises to become Wallenstein's steward. The only character entirely of Hindemith's invention, Tansur is Saturn, the outermost planet. Kepler's daughter, his narrow-minded brother Christoph and Susanna's foster-father, the baron Starhemberg, are not featured in the heavenly allegory.

An explicit albeit brief reference to the planetary model Kepler conceived in his *Cosmographic Mystery*, of orbits nested inside one another at distances defined by interpolated Platonic solids, occurs toward the end of act I, when Ulrich confesses his discontent with a master who speculates "as though a mind had puzzled over the cosmos playing with dice and tetrahedrons." More significant than this verbal allusion is a structural translation: acts I and II are music-dramatic instantiations of Kepler's first law.

Act I comprises eight musical segments distinguished by rehearsal cues. The through-composed music is framed by Tansur's song, which sounds with five stanzas at the act's opening and two at its close. The initial five stanzas, showing a Prague street scene in 1608, are followed in [2] by the dialogue at the same venue that lays the ground for the relationship between Wallenstein and Tansur; the final stanzas sound through an open window at the end of [8].

The stanzaic frame surrounds two sequences of musical numbers. In these, Kepler is shown in relation to four persons who were important to him during this period: his mother and young daughter in his private life, his employer and research assistant in the professional sphere. They are cross-aligned in intriguing ways. Both Emperor Rudolf and Kepler's daughter trust Kepler's understanding of the stars but are dismayed that it cannot give them more practical help; both Ulrich and mother Kepler are discontent with their daily lives and hope to use Kepler's insights for shortcuts to their own power. The two older persons, erstwhile authority figures who have now turned somewhat pathetic, are assailed by visions and temporarily out of contact with reality; the two younger ones are torn between their dependence on Kepler's guidance and their need for simpler explanations than those he is willing to admit.

The first sequence interweaves mother Kepler's visionary gift with Emperor Rudolf's experience of unreality and longing for a given order. The emperor is obsessed with the thought that as the supreme ruler he ought to understand, and if possible reign over, the universe. When a look through Kepler's telescope teaches him that the cosmos is much more complex than his mind could ever grasp, he suffers a fit of raving. Mother Kepler, too, is obsessed and confused about reality and the extent of her powers. Her insistence on defying conventions of acceptable behavior and following her path of superstition, exposed in her grave-digging aria and duet with Christoph [3], balances Emperor Rudolf's resigned recognition of his deranged state of mind in his conversation with Kepler [6]. In the heart piece of this sequence, the emperor's aria about the comfort expected of the cosmic order [4] and his subsequent interchange with Kepler [5], during which he grows increasingly upset, sound laced with mother Kepler's interjections, who "sees" and vicariously feels the emperor's distress at the cemetery in Württemberg, some 350 miles away.

The second sequence of musical numbers in this act links Kepler to the two persons who look up to him: his little daughter, who shares his grief and alleviates his loneliness [7], and his assistant, whose superficial ambition and lack of genuine dedication disappoint Kepler [8]. Both passages combine free recitative with arioso passages. The core of this second sequence is Kepler's lament over the recent deaths of his first wife and small son. In mood and intention, this second, very calm focal point is diametrically opposed to the violent scenes of dismay in Prague's imperial chamber and at the rural Württemberg cemetery.

Read metaphorically, the structure of act I appears as a visualization of Kepler's first law of planetary motion. Tansur's polemic, blending pseudo-astrological predictions of doom with political criticism, orbits around two focal points. One of them, the scene with the raving emperor, acts as the sun. The name Tansur is a transparent anagram for Saturn. Like the remotest among the planets known in Kepler's time, he has little contact with the other *dramatis personae* in this act as he travels his elliptical course around the center of gravity, the reigning emperor.

FIGURE 24: The orbit of opportunism and external forces

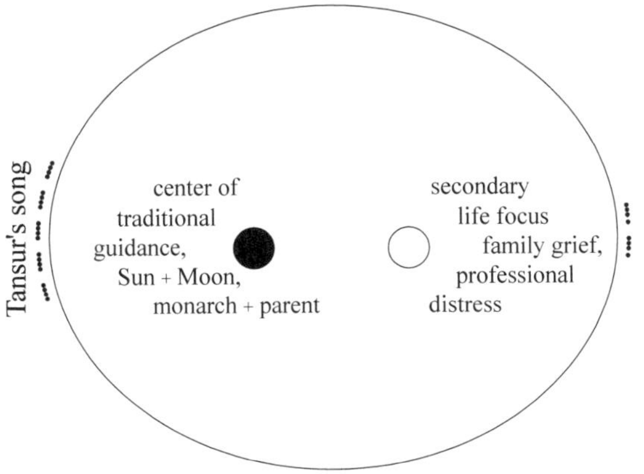

In act II as in act I, the music underscores a bracket that might otherwise escape notice. The opening scene in [9], in which Tansur accepts the task to pay off and chase the indigents living on the lot Wallenstein has chosen for his projected palace, is a rondo in the style of a march. A conspicuous passage from its second episode recurs, transposed and developed, in the final scene [16], in which Tansur is seen recruiting men for Wallenstein's next war. While it may appear as if Tansur was once again the character defining the framing segments, the determining force in both scenes is really Wallenstein. The general's actions so far seem almost as remote from the drama's main thread and concern as those of Tansur. His only contact with Kepler before act IV is the indirectly conveyed request for a birth horoscope, delivered by Ulrich who, as Mars, would indeed be the obvious mediator between Jupiter's and Earth's orbits.

The first focal theme in act II is musically captured in a choral fugue, a duet between Kepler and Hizler, and a passage for grand chorus and ensemble giving voice to the various reactions to the pastor's doctrinal autocracy (all united under a single cue, [11]. Correspondingly, the second focus comprises Susanna's monologue and duet with Kepler in [14], followed by Kepler's monologue and their second duet in [15]. The two climaxes are bridged by two further duets in which baron Starhemberg faces first Kepler, then Ulrich. Both address the two central issues: on the one hand, Kepler's religious conviction; on the other, Kepler's future marriage to a woman eager to share and support his goals.

FIGURE 25: Military might orbiting around Kepler's inner strength

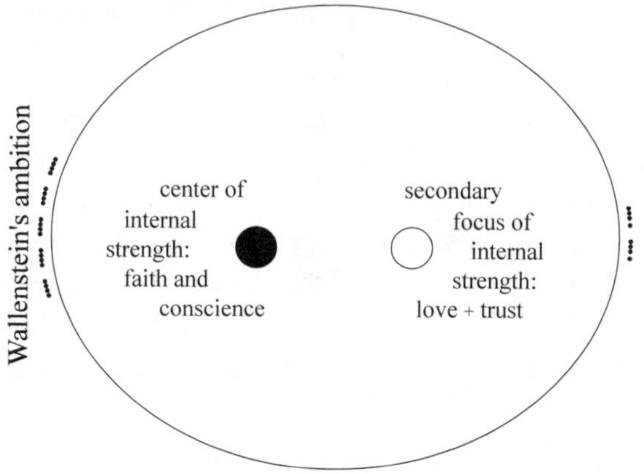

By setting the two segments in music of very different weight, Hindemith makes it clear that the first focus is once again paramount. Its emphasis on the conflict between individual conscience and decreed doctrine prepares the audience for noticing that the love scene, too, converges on declarations of creed and the conscientious objection to imposed agendas.

Act III abandons the "planetary-orbit scheme" of the two preceding acts: there is no frame created by an opening song or rondo and an act-closing reprise. Restricted to a multifaceted presentation of its two central themes, this act is at once more concentrated and more elaborate than the two preceding ones. The two themes, although ostensibly different in their dramatic presentation, are intimately related. As mentioned above, both address the shadowy side of human experience and perception, and both express themselves as circular motions around a single center. The cosmological and dramatic dimensions of this center are perfectly congruent. Just as the moon travels around planet earth, so do Kepler's mother, wife, and daughter move in the same spherical limit owing to Kepler's centripetal force, while the trial has everyone circling around the old woman.

FIGURE 26: Two rondos and a rondino in lunar circles

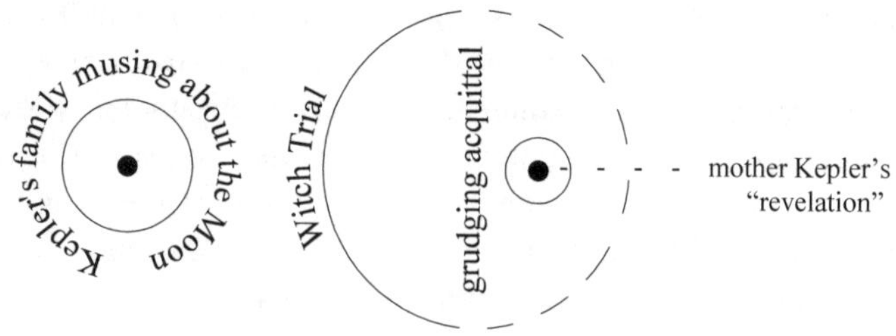

A Planet Unleashed

The two subsequent developments are designed in very different ways. The dramatic sequences as well as their musical settings suggests a trajectory that spins away from a center, never to return. To stay with the metaphor from astronomy: it seems that Hindemith is exploring what would happen were the sun's gravity ever momentarily surpassed by that of one of its larger planets. In this scenario, passing smaller planets might be first derailed and then flung out into endless space where they would be alone with themselves in a desolate situation. No longer part of a finite solar system, without warming solar energy or light they would move inexorably toward material and spiritual death.

Act IV describes the first part of this trajectory. The dramatic development is strictly unified here; the place is Wallenstein's newly built palace in Prague, the time, a single evening, possibly a mere half-hour segment of it. The central player is Wallenstein. As the leading general in a war that has already been raging for ten years and shows no sign of a letting up, he currently represents the center of power. Without him, the mighty emperor is forsaken, his hope to win his countries back for Catholicism lost. The sequence of encounters shows Wallenstein as a man who is much flattered and feared. He accepts the praise proffered him by influential men of State and Church, replying with grandiose promises; he responds to the soldier Ulrich's gawking admiration of his wealth with rage and a threat of hanging him for neglect of duty; he dodges Kepler's cautioning remarks about the limited value of horoscopes, overrides his uneasiness about a self-serving military adaptation of the idea of universal harmony, and humiliates the financially desperate scientist by inviting him to do the calculations for operations of conquest and suppression aimed against the Lutherans, i.e., people of Kepler's own creed.

Cosmologically speaking, the act bonds Earth (Kepler) with the two men whose allegorical planets orbit far from his own, at what was then thought to be the outer rim of the solar system: Jupiter and Saturn. Tansur, whose access to the shining Wallenstein's court rests on his willingness to be subservient and refrain from speaking his mind, will now be in charge of Kepler's pay and housing and thereby guide him toward a position that is wholly remote from that destined for Earth. Jupiter's light and energy in the form of Wallenstein's money and might exert a powerful attraction to all who come near him. When Kepler admits to himself that he is fascinated despite his better judgment, he enters on a tangential of his designated path and swerves outward, first toward Wallenstein and then, owing to the laws of gravitational pull, past him into the void.

Act V is at once bi- and tripartite with regard to its locale: Wallenstein's headquarter and the Kepler's home in Silesia, just as the Bavarian city of Regensburg where the electors meet with their emperor, are two venues on earth and presuppose human reality, whereas the assembly of the planets around the sun in the final three eighths of the act happens in the sky and in the realm of allegory. Relevant to the current analysis are only the earthly segments, which illustrate the consequence of Kepler's departure from his cosmic and spiritual path. The trajectory, now determined by Kepler's association with military and financial power and involving his descent into the world of warfare, develops its own momentum and continues relentlessly into the darkness of outer space.

In an initial scene (the protagonist is conspicuously absent), Hindemith shows that Kepler and his family have remained strangers in Wallenstein's Silesian headquarter. Susanna suffers from the enmity the surrounds them, which she considers proof of her failure to achieve her spiritual goals, while Kepler, she believes, is lost in a desperate search for an imagined "earthly harmony." This search leads him away from home so often that he loses contact with his family without therefore finding solace elsewhere. When he learns in the second scene that the scorn and hatred the country's princes and electors feel for Wallenstein will force the emperor to dismiss his general, Kepler sees the last support of his intellectual aims and his family's subsistence crumble. He recognizes that this blow (together with his overall exhaustion and a current sickness) will be his death.

In the imagery of the cosmographic metaphor I have been pursuing in this chapter, this is what happens: In the grip of intense worry about exposing his family to the "misery and famine" that the historical Kepler had called symbolic of planet earth, Hindemith's protagonist allows himself, in act IV, to be swept out of his unique orbit toward the gravitational center of the charismatic Jupiter. When Jupiter's ambition to establish a solar system of his own collapses, Earth must continue its outbound journey into the ever colder and darker expanse, where spiritual death will be followed by material destruction.

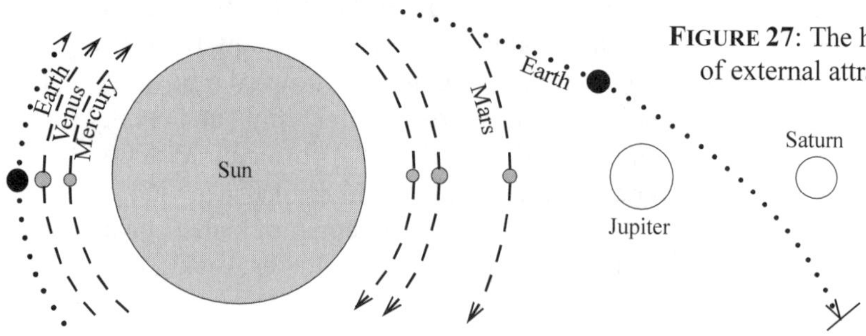

FIGURE 27: The hazard of external attraction

The Limits of the Logocentric World View

The 23rd-century World of the Mind seems in need of a reorientation of the kind Kepler's fellow astronomer Copernicus introduced to 16th-century Europe. The purifying spring that gave Castalia its name has lost many of its connotations. At ancient Delphi, it reminded those who drank of it to "know themselves," endowed them with the Apollinian gift of poetic elation, and promised them the muses' support. By contrast, third-millennium Castalia strives for a suppression of individuality and has banned most of the muses —certainly Melpomene, Thalia, Calliope, and Erato (the muses of tragedy, comedy, narrative fiction, and love poetry) as well as Clio (history) and possibly Terpsichore, the muse of dance. Through a sequence of "awakening" experiences, Knecht reintroduces history; his posthumous writings add poetry and narrative fiction, and shortly before his death, he grasps the ritual power of dance. He rediscovers the individual's inner voice, agrees to personal responsibility, and thereby readmits half of the excluded muses.

Artistic creation is considered anachronistic in Castalia primarily because of the value attributed to the glass bead game, claimed to be more than any particular art. To understand the philosophy of "playing" in this context, the most famous treatise on the subject, Johan Huizinga's *Homo ludens*, is less helpful—although Hesse did read it in the later stages of his work—than Schiller's *On the Aesthetic Education of Man*. Sections 13-15 of this 1795 treatise lead up to a passage that could serve as the main article defining Castalian existence were the Pedagogical Province ever to give itself a written constitution. For Schiller, humans are initially defined by two contrary forces, the sensuous drive and the formal (or rational) drive. In order for these to be harmonized, a third drive is needed, which will bring about "the union of reality with form, contingency with necessity, passivity with freedom." This synthesizing force is the play-drive. Schiller defines "play" as "everything which is neither subjectively nor objectively contingent, and yet imposes no kind of constraint either from within or from without." Rejecting reproaches that he overvalued "mere play," Schiller asserts "that it is precisely play and play alone, which of all man's states and conditions is the one that makes him whole and unfolds both sides of his nature at once." The passage climaxes in the famous declaration, "Man only plays when he is in the fullest sense of the word a human being, and he is only fully a human being when he plays."[9]

[9]Letter 15 in Friedrich Schiller, *On the Aesthetic Education of Man, in a Series of Letters*, edited, translated and commented by E.M. Wilkinson and L.A. Willoughby (Oxford: Clarendon, 1967), pp. 102-107.

This assertion, however, begs the question. Is the glass bead game player really the most fully realized human being? Does the Castalian enterprise, born some three centuries after the publication of Schiller's treatise, truly aim at the harmonization of the sensual and the rational forces in human nature? Schiller is not likely to have associated the word "sensual" with sublimated notions. A physician before he devoted himself fully to literature and philosophy, he had written his doctoral dissertation on "the correlation of a human being's animal nature with his spiritual-intellectual nature."[10] In his usage, "sensual" means much more than merely "aesthetic."

While the 23rd-century Castalia may have fallen short of the larger harmonization, Hesse's text stresses that Schiller's is indeed the ultimate goal. In fact, the chronicler credits two other 18th-century writers with having conceived poetic realizations of the idea that serve as models for Castalia. The first paragraph in his introduction ends with a mysterious reference to "our Province." Designed as a protected space in which the formation of free human beings can take place without any manipulation from State or Church and in which science and culture are sheltered from ideological and political instrumentalization, Castalia is a "Pedagogical Province" reminiscent of the one in *Wilhelm Meister's Journeyman Years*. The two models also share an emphasis on intellectual discipline and on an education that has music as its core. However, unlike Castalia, Goethe's institution stresses a higher harmony in two significant areas: in its training, where it balances intellectual achievement with skills in various crafts; and in the human ideal to be modeled, in whom respect for the hierarchy is to be matched with respect for each individual's self-realization.

Schiller did not explain what kind of "playing" he envisaged. For an explicit example we must turn to a younger contemporary, to whom Hesse alludes when he names among the forerunners of the glass bead game "the runes of Novalis's hallucinatory visions" (p. 16). In a novella fragment about an elite pedagogical scheme in which young intellectuals, "The Disciples at Sais," speculate about the world of nature, Novalis speaks of a cipher script capable of capturing the contents of the entire cultural tradition.[11]

Through his hints at Goethe's and Novalis's conceptions, Hesse alerts his readers in the first five pages of his text that something is awry in 23rd-century Castalia; and for the righting of this wrong, Joseph Knecht's life is deemed important enough to be lifted out of Castalia's prescribed anonymity.

[10]"Über den Zusammenhang der tierischen Natur des Menschen mit seiner geistigen" (1780).

[11]Martin Jacob Neeb, *Novalis, Christianity or Europe and The disciples at Sais*. A translation with critical introduction (University of Texas at Austin, Thesis, 1937). "Die Lehrlinge zu Sais" by Friedrich von Hardenberg (*pseud.* Novalis) was first published in 1802.

Copernican Mobilization: Toward a Holistic Concept

Contrary to what many Castalians think, the Province is part of a larger world in the same sense that planet earth is part of a solar system. Castalia's self-centeredness and self-importance can be likened to the geocentric model of the world, which laity and Church alike defended long and fiercely. To the lay person, a planet privileged with the emergence of *homo sapiens* could not but be the center of all that turns in the sky; for the Church, the birth on this earth of the man revered as the Son of God constitutes a divine statement about the spiritual focus of the universe. Copernicus and his supporters risked their lives when they declared that while mind and spirit in this solar system may indeed bloom on earth as nowhere else, nature nonetheless operates by different laws—laws the inhabitants of the earth would only understand and learn to use to their advantage once they accepted their own subordinate role in the cosmic scheme. By supplanting the geocentric with the heliocentric model, Copernicus highlighted a truth laity and Church would have preferred to ignore: that mind and spirit may be sentient life's most exquisite assets but are by no means nature's strongest forces.

In a symbolically analogous way, 23rd-century Castalia believes itself so privileged and refined that it has adopted the habit of thinking as little as possible about the remainder of the world. The colorful but disorderly realm of family and feeling, newspapers and amusements, politics and economic struggle, and of individual competition for success, status, and money makes Castalians shudder; the calm in the monasteries of the Roman Church strikes them as intellectually slow-paced and spiritually doctrinal. As a result of this attitude, relations with the outside world have dwindled to minimal polite contact. While antagonism is rare, mutual misconception and discomfort abound, and neither side strives toward a rapprochement. Castalia, which has after all grown from within the world-at-large and is constantly drawing on its material and human resources, is failing in its task of transforming it in turn. The only members who live up to its stated purpose are the teachers who leave the secluded province to work with students aiming to remain and succeed in the world. These teachers alone repay the sacrifices the world makes to keep Castalia alive; they contribute individually toward the much-needed harmonization. By contrast, the glass bead game, developed with the goal of universal unity—as a device aiming at the synthesis of all that is differentiated or even opposite—has neglected its ultimate pole.

Several Hesse scholars have pointed out that every objection that can be raised against Castalia is actually part of the narrative, in the form of a doubt or, later, an alarm felt by its protagonist. Joseph Knecht has a profound sense that the people of the Mind and the people of the sensual world are brothers.

Through his life and death, his career trajectory as a model Castalian and his startling transcendence of its territory and rules, he becomes a force of synthesis and re-integration. His quest for unity is driven by questions he asks already as a student in Waldzell: "why were these two world apparently unable to live in fraternal harmony, parallel and intertwined; why could an individual not cherish and unite both within himself"? (p.100)

Knecht's insight germinates as a result of his exposure to a unique array of mentors and debating partners. The venerable Music Master, who recognizes Castalia's dangers but finds ways to live with them in order to continue serving the high ideal, opens Knecht's eyes to the individual's choices and responsibilities on the ostensibly prescribed path. The one-sidedness epitomized in his brilliant but over-sensitive friend Fritz, and the isolation sought by another exceptional Castalian, the gifted Sinologist turned Daoist recluse known as the "Elder Brother," alert Knecht to elitism and escapism. The attacks the articulate guest student Designori launches against the parasitical life of a "priestly caste" and its "pack of spoon-fed eunuchs" (p. 96) make Knecht aware that mutual tolerance is precarious. Finally and most importantly, the criticism voiced by an influential Benedictine historian shape Knecht's insight into the limits and unforgivable omissions of the self-indulgent scholarly province. It is this insight, come to fruition in the mature *magister ludi*, that allows the chronicler to observe that Knecht has always been a genuine and faithful Castalian: he for one never forgot that Castalia is only a part—in fact: a very small part—of the world.

The integration of the World of the Mind with the world of nature is not achieved within the confines of the book; the harmonization of the intellectual-spiritual and the socio-political existences, toward which the whole work seems to strive, remains undepicted. For all his emancipated insight, Joseph Knecht is only shown to take the initial steps in this direction. But then Hesse's *Glass Bead Game* is ultimately—though one may for moments forget this—not about the self-perfection of an individual or his elite community, but about the idea of this game of games itself, the tool toward ultimate harmonization. As stated in the motto, "[n]othing is harder, yet nothing is more necessary, than to speak of certain things whose existence is neither demonstrable nor probable. The very fact that serious and conscientious men treat them as existing things brings them a step closer to existence and to the possibility of being born" (p. 9).

The effects of Knecht's integrative striving can only be gleaned indirectly, but the prime witness's testimony is persuasive. In the chronicler's early 25th century, Castalia not only continues to exist but seems to be thriving. During the roughly 200 years that have passed between Joseph Knecht's death and the time when the narrative is being composed, the World of the

Mind has apparently improved its integrative efforts toward the world-at-large. A tell-tale sign is the fact that the hierarchy's initial disapproval of Magister Knecht's "defection" has given way to a recognition that his decision represented in fact obedience to the true spirit of the Game.

This is not to say that the final integration of the World of the Mind and the world of nature has been achieved even then. The ultimate harmonization may forever lie in the future—it may prove to be a goal never fully to be reached. At any rate, the changes for which Knecht hoped and strove can only be effected by processes of history—processes that Knecht hoped to set in motion through the example of his own life.

Knecht's action is threefold, comprising thought, speech, and deed. In the course of three decades of doubt, self-questioning, observation, and the courage to assume responsibility for his conformance or resistence to the protective hierarchy, he opens himself for an understanding that his fellow Castalians prefer to avoid. Risking rejection and ridicule, he communicates his misgivings to those who least wish but most need to hear about them. Finally, he leaves a life for which he was singularly trained, in order to be true to his own thought and word but also to underscore the urgency of his plea with the highest possible stake. Having taken this step, he meets the end of his life—an end that may seem rather abrupt for the taste of readers relishing biographies and novels, but is consistent with his role in the larger narrative of which this life story is but one component. As Theodore Ziolkowski explains: "It would be pointless for Knecht's life to go on; he could hope to accomplish no more than has been accomplished."[12] The nameless narrator, writing many generations after Knecht's death, is the living voice of a Castalia that has taken the message to heart and has—no doubt painstakingly and over several generations—instituted some of the remedies for which Knecht strove.

For 25th-century Castalia with its continuing world of hierarchy and elite, schools, archives, and game village, the exceptional *magister ludi* Joseph Knecht served as a prophet, forerunner, and instigator. This is not, however, his only role in Hesse's work. He is not merely the protagonist in the central life story, but also the secondary subject matter in a narrative whose leading theme is the glass bead game. In this context, the sequence and effect of his individual experiences are not merely embedded in, but also symbolic of, the sequence and desired final outcome of the game's harmonization process.

[12]Ziolkowski, *The Novels of Hermann Hesse: A Study in Theme and Structure* (Princeton, NJ: Princeton University Press, 1965), p. 336.

Five-Step Harmonization in Life and Game

Synthesis is vital to the idea of the glass bead game. As the chronicler tells it, its development knew four distinct stages. In the first, different aspects of music were abstracted so that their essence could be captured in a game with glass beads strung on strings. In step 2, a formulaic language was developed that could be adapted to analogous exercises in other fields, leading among them mathematics. Stage 3 brought the move toward universalization, attained on the basis of a unique language of symbols and ciphers that allowed to establish correspondences between the insights of the various fields of culture and knowledge. In the fourth stage, the intellectual-aesthetic content was supplemented with a form of contemplative absorption, introduced to safeguard the game from becoming a mere display of virtuosity. Meditation became an essential component both of the game in its private and ceremonial versions and in the individual Castalian's life. It fostered maturation and self-possession in this otherwise overly intellectualized ambience and, since it so obviously reached beyond the realm of pure mind and reason, deceived its practitioners into believing that they had achieved the highest synthesis.

If nothing else, the number of stages—four—ought to alert readers. As previous chapters in this study have demonstrated, the magic number in Hesse's *Glass Bead Game* is invariably FIVE. It is no coincidence that the four "movements" of the discursive glass bead game analyzed earlier illuminate the subject matter captured in its themes from many angles but do not yet achieve the ultimate harmonization expected at the conclusion of a successful game. Just as this discursive game will find inner completion only in the fifth movement—a movement that takes the unexpected form of a collection of poems written clandestinely, in disregard for one of Castalia's ground rules—so also will the glass bead game as idea and tool have to strive for a fifth stage in its developmental process. This highest synthesis will have to be one that unites the World of the Mind and the world of nature.

Joseph Knecht's life is a symbol of this gradual integration not thanks to the official education he receives or owing to a specific and unique talent, but as a result of a quite different character-shaping force: his attentiveness to his mentors and debating partners. Of these persons, whose thinking and behavior influences Knecht in ways so profound that they overrule his training as a self-contained intellectual and his indoctrination as a submissive member of a powerful hierarchy, there are five. Each of them leaves him a more thoughtful person. If the influence of one—the guest student of his Waldzell years—is initially inconclusive, it ripens to full fruition in the scenes preceding Knecht's death.

The cornerstones of Knecht's character education are his two mentors, the Music Master and Father Jacobus. They are "cornerstones" both in that theirs is the first and the last significant force of influence in Knecht's life, and in that both have found peace in their chosen roles as members of an order but nonetheless set Knecht free to address his own doubts as he must. As Adrian Hsia shows, the Music Master combines a few traits of a Daoist with many characteristics of a Confucian sage. His incomparable humility as well as his transfiguration during his final months—his enchanted withdrawal into a serene, radiant silence, "away from words and toward music, away from ideas and toward unity" (p. 257)—can be called Daoist. But, as Hsia affirms, "it is inconceivable that a Daoist of the schools of Laozi or Zhuangzi be a civil servant, and what is more, one as efficient as the Music Master. A Daoist would never hold one of the highest offices. The way in which the Music Master furthers Joseph Knecht's career is positively un-Daoist."[13] Equally un-Daoist according to Hsia is the feeling the older man expresses on the occasion of Knecht's admission into the Order, when he says, "it is as if I had a son who will stand in my stead" (p. 143). An emphasis on the ties between father and son, teacher and pupil, and a man's duty to the body within which he is placed to serve is characteristic of Confucian virtues. Similarly, Father Jacobus is not a monk who lives absorbed in the service to his God at the exclusion of all else, but rather the monastic variant of a worldly sage. Though under vows of poverty, celibacy, and obedience like any member of an order, he has never lost his profound interest in and contact with the world. In words quoted from his model, the historian Jacob Burckhardt (1818-1897),[14] Father Jacobus conveys to his Castalian visitor an appreciation of the individual's role in times of upheaval, of the importance of the place any being or event occupies within socio-political history. If the Music Master is living proof of the responsibility the individual owes to his allotted place in whatever social structure, Father Jacobus stresses devotion to a responsibility to life in all its aspects.

The two men who influence Knecht most profoundly during his years of free study are his closest friend, the glass-bead-game genius Fritz Tegularius, and his instructor in the *I Ching*, a Sinologist nick-named "Elder Brother." Tegularius is capable of dazzling work when left to his own whim,

[13]Adrian Hsia, "I Ging und das Glasperlenspiel Hesses," in *text und kritik* 10/11 *Hermann Hesse* (1977) : 53-66 [59].

[14]Knecht's circular ends with a postscript quotation attributed to the Benedictine historian (cf. "Times of terror and deepest misery...," p. 363) which is a verbatim quotation from Burckhardt's lecture on the revolutionary age, published in *Historische Fragmente aus dem Nachlass* (Berlin: Deutsche Verlagsanstalt, 1929), vol. 7, p. 426.

but unfit for service or responsible contribution to a larger project owing to his precarious nervous condition. Similarly, Elder Brother exceeds everyone (including the visiting scholars from China) in his understanding of the Chinese language and culture but, alienated by the "antimystical and declaredly Confucian spirit of the Chinese Department" (p. 127), has withdrawn from society to a Chinese hermitage complete with bamboo grove, goldfish pond, and a provision of ancient scrolls to be copied. Here he has been living in retreat for a quarter century, serving no-one but his science and his quest for serenity.

Both men are exceedingly talented Castalians whose personal preferences or weaknesses make them into outsiders. While they appear as embodiments of the two symbolically cast wisdom traditions that shape Knecht's thinking—the cultural tradition with its metaphor, the glass bead game, and the spiritual tradition with its tool, the *Book of Changes*—the two men also alert Knecht to the flaws of a society that causes their isolation and deprives it of the use of their talents. Privileged as their idiosyncratic and shielded existence may seem, it can be read as a symptom of Castalia's problematic detachment from life's non-intellectual conditions and requirements.

Hesse's choice of the men who influence Knecht's thinking is ingenious, in that he juxtaposes two "servants" with two individualists and, cross-wise, two representatives of ancient Chinese attitudes with two embodiments of a modern (or future) Western stance. One of the two older men, the obedient and successful Castalian *magister musicae*, urges Knecht to stay connected with the deeper layers of transindividual freedom and serenity, while his counterpart, the Benedictine father, teaches Knecht the importance of change in its historical context. One of the two younger men, the overly sensitive and nervous intellectual Fritz, cannot endure any deviance from his routines and has lost all touch with nature and a simpler life, while the other, Elder Brother, will not submit to the demands of a socio-ethical teaching but only follow his need to blend into nature. Among the "Westerners," the glass bead game player Fritz suffices himself as an engrossed intellectual, while the monk and scholar of modern history is preoccupied for the larger world. Among the "Chinese," the self-serving Daoist recluse lives detached from all social bondage at the spring of a literal and figurative river of purification, while the "Confucian" Music Master exercises the many tasks of a leader called to one of the highest offices in the Castalian hierarchy, thus exerting himself in the service of the whole.

The fifth man of crucial influence on Knecht's life is the guest student Plinio Designori who subsequently attempts but fails to synthesize Castalian ideals with the requirements of the outside world. In his protracted debate with this critic of Castalia, young Joseph feels first doubts about the Order and its elitism but remains dialectically sovereign and ostensibly unscarred. Only when his own awakening has reached the point at which he feels he must change his life, is this unachieved attempt at harmonization taken up in a new spirit.

Insofar as Joseph Knecht is not only the hero of the Castalian chronicler's "secular hagiography," but a personalized symbol of the *Glass Bead Game*, his inner development toward an ever more encompassing integration of viewpoints mirrors the game's quest for universal harmonization.

FIGURE 28: The 5-step process of integration in the life of Joseph Knecht

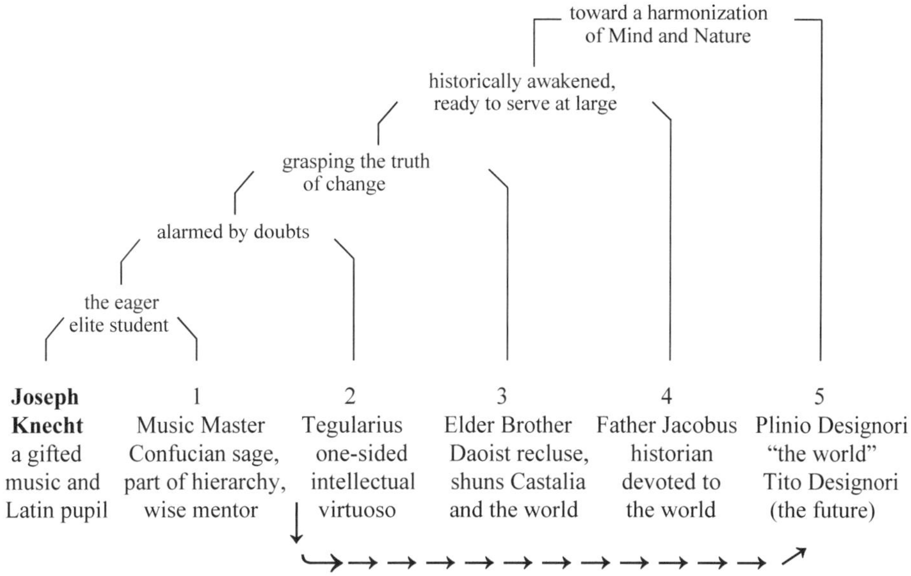

FIGURE 29: The 5-step process of harmonization in the glass bead game

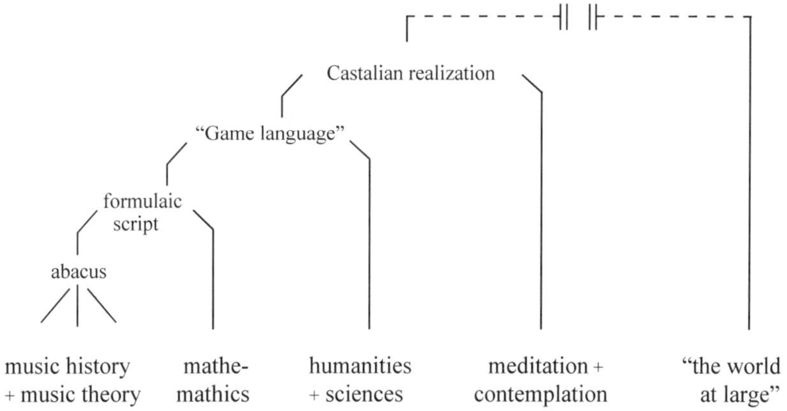

Knecht's abrupt and at first glance narratively incomprehensible death is the only conceivable ending of the Life within Hesse's *Glass Bead Game* for two additional reasons. Readers invested in the extraordinary *magister ludi*'s biography may regret not seeing him achieve his private harmonization of the two worlds, both in his own person, as a Castalian living and serving in the world, and through the successful molding of a promising but pampered youngster into a human being who combines the best of mind and nature. But even with the greatest conceivable success, he would have achieved no more—quite possibly considerably less—with regard to the larger issue at hand, the necessary Castalian step toward the fifth level of synthesis, then he has at the time of his drowning.

For readers of Hesse's compound work, any unproblematic extension of Knecht's life in more or less mundane surroundings would actually have turned out deceptive, in that it would have tempted them to mistake the signifier for the signified, the symbol for that to which it points. With a long, well-rounded and fully satisfying life story, Knecht would have lured us into regarding his person as the center of Hesse's message. The material that traces the larger question—the chronicler's introduction, the many reflections about Castalia's relationship to the world, the conjectural biographies, and the poems—would then have appeared as no more than curious appendices or somewhat disruptive digressions to a hero's tale or hagiography.

Knecht's vexing death jolts readers out of this illusion. It confronts them with the need to make sense of the narrative whole, to sense correspondences and discover patterns of symbolic relationships that might otherwise have remained hidden.

Part V
METAPHYSICS

PLATE 5: Philosophy in the center of the seven liberal arts. Illustration from the Abbess Herrad von Landsberg's *Hortus deliciarum* (folio 32), late 12th century. Reprinted from Gérard Cames, *Allégories et symboles dans l'hortus deliciarum* (Leiden : Brill, 1971), plate IV.

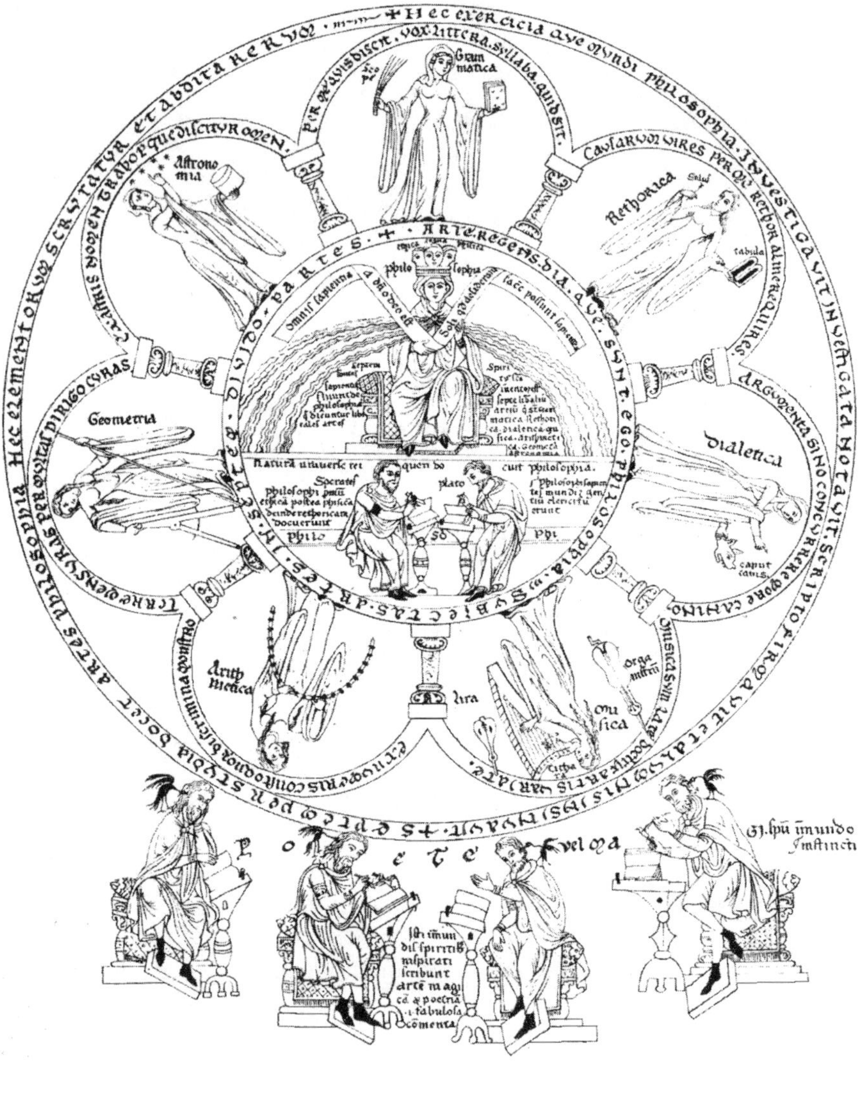

The Musical Nature of the World-Soul

A creation myth of Maidu Indians in Northern California describes how Earthmaker, having generated by the sheer force of his mind a first lump of clay, *sings* over it in order to make it grow and then thrusts it into space. In a creation poem of the Métis people (indigenous of Canada) as narrated by their female Elders, the world came into being through the *song* of "the old lady." Even the initial words of the Gospel of John could be rephrased similarly without losing their essential message: "In the beginning was the song, and the song was with God, and the song was God. . . . And the song became flesh, and dwelt among us." The divine Logos might usefully be imagined as singing, expressing itself in the heightened form of poetic utterance rather than just another version of our mundane kind of communication. To imagine hearing God singing the Word provides a new appreciation of the depths of love from which God desires that there be a counterpart in whom He might see Himself reflected. This is certainly how children like to dream: in *The Magician's Nephew*, C.S. Lewis tells of the creation (and subsequent initial corruption) of a world called Narnia, which the lion Aslan, a representation of God, "sings into being."

A more sophisticated myth is told by Zhuangzi (Chuang Tzu), an early Daoist poet. He describes how the Yellow Emperor has music performed in the wilds around a lake. This music, "performed by man, tuned to Heaven, played with ritual principle, and established in Great Purity," confuses and overwhelms listeners. It is perfect music, which must "respond to the needs of man, accord with the reason of Heaven, proceed by the Five Virtues, and blend with spontaneity; only then can it bring order to the four seasons and bestow a final harmony upon the ten thousand things." Asked what this music signifies, the emperor says: "When the heavenly mechanism is not put into action and yet the five vital organs are all complete, this may be called the music of Heaven. Wordless, it delights the mind. Therefore the lord of Yen sang its praises thus: 'Listen—you do not hear its sound; look—you do not see its form.' "[1]

In the philosophical circles of ancient Greece, Plato was the most evocative proponent of a creation myth. His dialogue *Timaeus*, dated 360 B.C.E., develops a cosmology and corresponding natural philosophy. It describes the

[1] *The Complete Works of Chuang Tzu*, trans. by Burton Watson (New York: Columbia University Press, 1968), pp. 156-158.

creation of the large and the small—the cosmos in general and the human being in particular—as beginning with the creation of the soul. This soul is at once the World-Soul and the stuff of individual souls. Plato conceives it as a synthesis and intermediary between the Essence of the universe and the universe itself. It was created prior to the material world and thus continues to enjoy precedence in value over it.

Since Plato's detailed description of how the demiurge fashioned the World-Soul has striking relevance for harmonic thinking (harmony once again understood as concerned with proportion, not as the modern concept of chord relationships), it is worth quoting the relevant lines in full. Note that the bracketed numbers in Benjamin Jowett's translation are his own numerical interpretations of the text's verbal descriptions. They are not found in the Greek original.

Once the demiurge has fused the indivisible, unchangeable Essence with the divisible matter of the physical world as well as a third component that He has concocted by mixing the other two, he begins the "harmonizing" work.

> He again divided this whole into as many portions as was fitting, each portion being a compound of the same, the other, and the essence. And he proceeded to divide after this manner: First of all, he took away one part of the whole [1], and then he separated a second part which was double the first [2], and then he took away a third part which was half as much again as the second and three times as much as the first [3], and then he took a fourth part which was twice as much as the second [4], and a fifth part which was three times the third [9], and a sixth part which was eight times the first [8], and a seventh part which was twenty-seven times the first [27].

The proportionally sized lumps the demiurge sets aside are listed in an order that, with its reversal of direction and strange skips (. . . 4, 9, 8, 27), appears puzzling to non-initiates but alerts those in the know. In fact, the numbers recall a numerical instance of the letter *lambda* that was popular with the Pythagoreans, in which the powers of 2 are juxtaposed with the powers of 3 in two limbs joined at the central, unifying "one":

$$
\begin{array}{ccccccc}
 & & & 1 & & & \\
 & & 2 & & 3 & & \\
 & 4 & & & & 9 & \\
8 & & & & & & 27
\end{array}
$$

The *Timaeus* continues with a passage that seems even more mysterious, outlining the next steps in the demiurge's work:

> After this he filled up the double and the triple intervals cutting off yet other portions from the mixture and placing them in the intervals, so that in each interval there were two kinds of means, the one exceeding and exceeded by equal parts of its extremes, the other being that kind of mean which exceeds and is exceeded by an equal number.

Let me unpack this. The demiurge filled the "double intervals" and the "triple intervals" (i.e., the spaces between 1, 2, 4, and 8 on the one hand and those between 1, 3, 9, and 27 on the other), each with two values. These values are the *harmonic mean* (described here as "exceeding and exceeded by equal parts of its extremes") and the *arithmetic mean* ("that kind of mean which exceeds and is exceeded by an equal number"). The harmonic means to be fitted into the spaces of this *lambda*'s two limbs are:

for 1-2-4-8: $1 - \mathbf{1\frac{1}{3}} - 2 - \mathbf{2\frac{2}{3}} - 4 - \mathbf{5\frac{1}{3}} - 8$,
for 1-3-9-27: $1 - \mathbf{1\frac{1}{2}} - 3 - \mathbf{4\frac{1}{2}} - 9 - \mathbf{13\frac{1}{2}} - 27$.

The values for the arithmetic means are:

for 1-2-4-8: $1 - \mathbf{1\frac{1}{2}} - 2 - \mathbf{3} - 4 - \mathbf{6} - 8$,
for 1-3-9-27 $1 - \mathbf{2} - 3 - \mathbf{6} - 9 - \mathbf{18} - 27$.

The fact that the entire passage is alternately known as Plato's "Creation of the World-Soul" or "the Timaeus Scale" gives away the solution to the riddle. Recent scholarship believes the text to hide a secret teaching most probably originating in 4th-century B.C.E. Pythagoreanism. The numerical play is then understood to describe the nature of the World-Soul as made up of consonant parts such as humans find manifested in a musical scale. Assuming a fundamental value 1 = C, a merger of the sequences results in a numerical series that translates, first into fractions representing vibrational ratios, then into tones:

1	$1\frac{1}{3}$	$1\frac{1}{2}$	2	$2\frac{2}{3}$	3	4	$4\frac{1}{2}$	$5\frac{1}{3}$	6	8	9	$13\frac{1}{2}$	18	27
1	$4/3$	$3/2$	2	$8/3$	3	4	$9/2$	$16/3$	6	8	9	$27/2$	18	27
C	F	G	C	F	G	C	D	F	G	C	D	A	D	A.

The World-Soul thus sounds in a combination of the tones many peoples and cultures of this world consider supreme: the Pentatonic set C–D–F–G–A.

The text's last portion is by far the most esoteric. Plato concludes:

> Where there were intervals of $\frac{3}{2}$ and of $\frac{4}{3}$ and of $\frac{9}{8}$, made by the connecting terms in the former intervals, he filled up all the intervals of $\frac{4}{3}$ with the interval of $\frac{9}{8}$, leaving a fraction over; and the interval which this fraction expressed was in the ratio of 256 to 243. And thus the whole mixture out of which he cut these portions was all exhausted.

If we read the fractions as indicators of ratios describing musical relationships, the "intervals of $\frac{3}{2}$ and $\frac{4}{3}$ and of $\frac{9}{8}$" correspond to fifths, fourths, and whole tones. By filling these with further intervals of $\frac{9}{8}$, the demiurge inserted more whole tones into the "gaps" of the fourths and fifths. Considering the first four values described above, C–F–G–C, the creator of the World-Soul apparently fitted D and E into the first fourth, A and B into the second. As Plato has it, these whole tones would all have been of the size corresponding to the ratio of the 9th to the 8th partial. As a result, the second whole tone in each tetrachord (E and B in this example) sounded slightly sharp ("the Pythagorean third") compared to the natural partial. Consequently, the semitones E–F and B–C (the "fractions left over") were smaller than they would be in the natural overtone series—their numerical value being precisely the striking ratio Plato mentions for these remainders.[2]

The operations mentioned in the third component of Plato's passage thus complement the Pentatonic scale to a diatonic seven-tone scale. Under the modern assumption that the fractions represent vibrational frequencies (i.e., partials), there results the ascending major scale—on C, in the given example. In the ancient Greek system, however, scales were naturally descending, as results also from Plato's passage if all fractions are interpreted as representing the relative length of the sounding bodies (strings or air columns). The descending scale with two tetrachords each comprising two whole tones followed by a semitone is E–D–C–B–A–G–F–E.[3] This was the principal scale in the ancient Greek system, and E, the era's principal tone.

Plato envisaged the World-Soul as a scale serving as basis of an entire musical universe. Like the Pythagoreans, he was obviously convinced that the cosmos and all nature are built according to the normative proportions known from music. Consequently, he ascribed to the spiritual force constituting this cosmos a structure that reflects these superior properties. The fact that he did not express this plainly but veiled it in fairly enigmatic language proves that he knew about the significance the Pythagoreans accorded this secret—and sacred—knowledge and respected the power of its symbols.

[2]Mathematically inclined readers may wish to check this: Within the tetrachord C–F or 1–$^4/_3$, with D sounding at its natural $^9/_8$, the second whole tone, E, now equals $^9/_8 \times {}^9/_8 = {}^{81}/_{64}$ ($^1/_{64}$ more than its natural value within the series of partials, which is $^5/_4 = {}^{80}/_{64}$). What remains between this E and the subsequent F is, then, $^4/_3 \div {}^{81}/_{64} = {}^{256}/_{243}$.

[3]Until the medieval church turned the scales upside down and reassigned the Greek names, this scale would have been called "Dorian." For more on this subject matter, see, e.g., Rudolf Wittkower, *Architectural Principles in the Age of Humanism* (New York: St. Martin's Press, 1988) and Ian H. Johnston, *Measured Tones: The Interplay of Physics and Music* (Bristol, PA: Institute of Physics Pub., 2002).

Kepler considered the *Timaeus* one of the most significant texts ever written. As he wrote in *Harmonices mundi*, he understood Plato's description of the creation of the World-Soul as "a kind of commentary on the first chapter of Genesis."[4] He felt that the *Timaeus* contained the seeds for making the study of planetary motion more precise.

Kepler's initial reasons for rejecting the Ptolemaic model of planetary motion, whose geocentrism the Church regarded as a matter of orthodoxy, in favor of the Copernican, heliocentric one, were not empirical but based on theoretical insights and metaphysical assumptions. He believed that God must have created the world to reflect divine aesthetics. He saw it as consisting of "archetypes," to which he would resort whenever physical arguments failed to explain what he was trying to understand. While the term seems Jungian, the concept actually goes back to Greek antiquity. For Aristotle, the proper way to explain the behavior of physical objects was to deduce it from a consideration of their essences. An object's mathematical coordinates, Aristotle argued, are accidental rather than essential properties. Kepler, concurring with Aristotle in principle but refuting the latter claim, developed a metaphysics under which the mathematical properties of an object *are* its essential properties. God's aesthetics, he was convinced, is essentially geometrical in nature. Once one has analyzed the details of the cosmic elements that are accessible to our measuring instruments, one can infer the general essence—for example, whether the cosmos is Copernican or Ptolemaic. Conversely, having grasped the structural features, one can understand God's aesthetic plan.

Kepler's work is a rigorous attempt empirically to test the validity of his metaphysical world view. He believed, as had Pythagoras 2160 years earlier, that the created world is permeated by harmony (all measurable aspects are designed in ideal proportions to one another) and that these proportions constitute the transcendentals: the Good, the True, and the Beautiful.

Not long after Kepler, the German philosopher Gottfried Wilhelm Leibniz (1646-1716) was strongly interested in harmony, particularly in its universal aspect. Like so many other thinkers from Pythagoras to Kepler, he derived the idea of a world defined by consonances from observations about the sonic manifestation of ideal proportion, calling music "the imitation of that universal harmony which God has put into the world." When he declares that "music is the hidden arithmetical exercise of a mind that does not know it is calculating," he seems to suggest that the subconscious realization of mathematical proportions is the ultimate *cause* of music's sensuous effect.

[4]*The Harmony of the World*, trans/ed. E.J. Aiton, A.M. Duncan, J.V. Field (Philadelphia, PA: The American Philosophical Society, 1997), p. 301.

Harmony is a central concern of Leibniz's philosophy, and a symbol in what he called "prestabilized harmony." In his philosophy, the universe is composed of countless conscious centers of spiritual force or energy, known as monads. Each monad represents an individual microcosm that mirrors the universe in varying degrees of perfection while developing independently of all other monads. The universe as the sum of these monads is the harmonious result of a divine plan.

Arthur Schopenhauer (1788-1860), writing his magnum opus, *The World as Will and Representation*, exactly 200 years after Kepler's *Harmonices Mundi*, adopted Plato's doctrine of ideas as the permanent forms of reality underlying all phenomena, but regarded them as standing between the one will and its differentiated manifestation in the world of sense—as intermediaries, not ultimates. Schopenhauer believed that Platonic ideas are manifested in works of representational art, whereas ultimate reality can only be articulated in the one art that is inherently non-representational: music. Accordingly, he emended Leibniz's sentence to read, "music is the hidden metaphysical exercise of a mind that does not know it is philosophizing."

The two most outstanding modern thinkers on the metaphysical aspects of harmony are the German lawyer and philologist, Albert von Thimus (1806-1878) and the Swiss art historian and music theorist Hans Kayser (1891-1964). Thimus, who had studied Chinese, Arabic, cuneiform writings, and hieroglyphics, believed that the fundamentals of music—above all, intervallic proportions—were the basis of teaching in the ancient Chinese, Hebrew, Egyptian and Greek civilizations. In his monumental work, *Die harmonikale Symbolik des Altertums*, he attempts to establish a symbolic expression of these fundamentals as a formulating principle of ancient cosmogonic theories and metaphysical views both in the West and worldwide.[5] Kayser spent his life in the service of an exhaustive compilation and synthesis of harmonic facts from all areas, with the aim to corroborate the breadth of Kepler's research in "world harmony." His unique achievements are "aural images" of crystals, plants, and animals (physical features expressed as ratios and then translated into tones); the interpretation of the *lambdoma* matrix as symbolic of various religions as well as mystic approaches to the divine; and the development of a "harmonic ethics." Albert Einstein, who famously expressed his faith that the principle of the universe is "beautiful and simple," would have applauded.

[5]A. von Thimus, *Die harmonikale Symbolik des Altertums* (Hildesheim: Olms, 1988 [1868]). For a helpful introduction to and critique of this difficult work, see Leopold Spitzer, *Die Harmonikale Symbolik des A. von Thimus: Ein Beitrag zur Geschichte der Harmonik* (Vienna: Lafite, 1978).

Kepler's Religious Conviction in Life and Opera

Questions of metaphysics loomed large in the historical Kepler's life. He combined a pious Christian's religious searching with a pensive outlook on the transcendental dimensions of science and art, communal and individual livelihood. The foundation for this general attitude was no doubt laid by the particular kind of schooling he underwent. During the years corresponding to today's primary-school education, he spent one year in a program that established literacy (a "school for German reading and writing") followed by three years at the so-called Latin School. As his biographers note, upon graduating from this school pupils had mastered Latin in reading, writing, and speaking, had studied the catechism and the Bible, and had received training in creative writing as well as in music. Literature read by the 12-year-old seniors included Cicero, Terentius, and Esop's fables.

This would have ended formal schooling for a boy from a family of limited means, had Kepler not passed the State Examination. Despite its secular name, this examination selected the most gifted students not with the aim of opening doors for them to just any academic career. Instead, the chosen few (about 25 each year in the state of Württemberg) were destined as future Lutheran clergy. If the boys received an excellent liberal arts education, this was owed to the image the Lutheran Church had of its clergy as spiritual and intellectual leaders. Thus at 12, Kepler entered one of the "monastery schools"—institutions housed in former cloisters and convents that had been transformed for the education of the Lutheran progeny. Pupils lived in monk's halls, began their days year-round at 5 a.m. with the singing of psalms, wore black cloaks, were forbidden to leave the premises individually, received all of their teaching in Latin, and were even expected to speak Latin among themselves. The lower of the two schools they had to pass through offered a two-year curriculum with literature, rhetoric, dialectic, and Greek in addition to the subjects continued from the Latin School: Latin, theology, creative writing, and music; the three-year-long higher school added arithmetic and astronomy.[6]

For 21st-century parents who fear that such a demanding education must stifle a teenager's non-academic potential, the amount of creativity and independent thinking the best of them showed may come as a surprise. Kepler's self-assessment is full of examples.[7] Many of the young Kepler's attempts

[6]For an excellent presentation of Kepler's early experiences—intellectual as well as spiritual and emotional—see Mechthild Lemcke, *Johannes Kepler*, pp. 16-28.

[7]See his "Selbstcharakteristik" in Franz Hammer, ed., *Johannes Kepler: Selbstzeugnisse* (Stuttgart-Bad Cannstatt: F. Frommann, 1971), pp. 16-30 [16-17].

at poetry evidently went far beyond required homework. The boy thrived on challenges and imposed them on himself whenever he could. This included formal demands like those posed by acrostic and anagrammatic poems or, later, by dithyrambs and Pindarian odes, as well as the provocation of subject matters unlikely for poetic presentation, such as the immobility of the sun, the genesis of rivers, the view from Atlas upon mists, etc. Another area in which he combined poetic skill with research was allegory, which the young Kepler pursued into minute details.

Kepler's spiritual sensitivity was equally astounding. Even before age 12, when he formally embarked on the path toward Lutheran theology, he began pondering the nature of the Eucharist. As he later told one of his mentors at Tübingen, he had heard a Lutheran deacon deliver a sermon about St. Paul's letter to the Romans in which the preacher took pains to prove the Calvinists wrong, who held that Christ's body and blood were present at communion *spiritually* rather than materially. Feeling confused, the boy went home, read the biblical text for himself and concluded that to him, the Calvinists seemed in fact very convincing. During his monastery-school years, this question was frequently addressed, along with a warning of the Calvinist misrepresentation. On such occasions, Kepler invariably withdrew into self-questioning. He understood that his interpretation was problematic in view of the reigning orthodoxy, but felt bound to be true to his conscience.

Another issue that began to haunt him during his early teens was the doctrine of predestination. Acknowledging that his conscience demanded that he study the matter before coming to a conclusion, the 13-year-old wrote to the Tübingen seminary asking to be mailed an essay on the subject. Initially assured that the human will was unfree in the Lutheran sense, he now leaned toward the opposite view. Moreover, a meditation on God's grace convinced him that pagans could not be destined to outright damnation, as the Lutheran Church taught. Kepler's independence of thinking defined his whole life; he always granted opposing views his full attention and respect before possibly deciding against them. As can be imagined, the seriousness of his religious questioning provided fellow pupils with ample opportunity for mockery.

When the 19-year-old Kepler, inspired by his undergraduate studies in the liberal arts and proud of his master's degree, began his professional studies in Lutheran theology, his sense of unintentional spiritual nonconformism returned with a vengeance. Learning that the Calvinists were not alone in opposing the Lutheran view of the Eucharist by stating that Christ's body and blood were present "through the *logos*," but that this opinion was shared by the Jesuits and validated by many of the church fathers, Kepler began to feel that it was the Lutheran view that seemed aberrant. Dismayed by his inner conflict, he sought out a mentor in the hope of finding understanding for his reservations about the absoluteness of the Lutheran doctrine. Yet when he argued in the spirit of original Protestantism that every believer must understand the Bible as it was given to him or her individually, he had to learn that this kind of freedom had not outlasted the heady days of the Reformation. The Lutheran Church, in Württemberg as in many other parts of reformed Europe, permitted no deviation from the Augsburg Confession of 1577, in which both the creed in general and the doctrine of the Eucharist in particular were spelled out in no uncertain terms. Every Lutheran clergyman had to sign an oath stating his agreement with this formula. As Kepler later wrote in his self-assessment, his call as a mathematician to Graz saved him from having to face the frightening conflict between his conscience and his Church.

While stances vis-à-vis the Eucharist, the concept of predestination, and the universality of divine grace are the most obvious proofs of Kepler's need to follow his inner voice, the questions on which he focused in his scientific work show an equally independent spirit. He would not discredit Copernicus's heliocentric model or Galileo's observation of mountains and craters on the moon just because the Church wanted the earth to be central and the heavenly bodies to be perfect with smooth surfaces; yet he was eager to prove that the particular way in which the solar system was laid out, as well as the motions of its parts in relation to one another, sang the praise of God the supreme and beneficent Creator. As his summary remarks at the end of *Harmonices mundi* attest, he considered his laws of planetary motion as well as his calculations of the orbital speeds mere means toward the greater end: they allowed Kepler the humble scientist to make known to the world the skill, artfulness, and perfect "music" in the divine Creator's work.

In Hindemith's opera, all these topics are addressed through a variety of dramatic, verbal, and musical means. Central in all respects is Kepler's conflict between his conscience and the expected religious conformity. At two points in the plot, a Lutheran pastor derides Kepler for his departure from Lutheran doctrine or, more particularly, for his deviant interpretation of the Eucharist. The music Hindemith invents for the encounter with Pastor Hizler from Linz (act II, at [11 C-I]) recurs in the death scene when the pastor from Regensburg attempts a last-minute conversion to "the right view" (act V, at [27 D-K]).

In the first of the two related scenes, the gist of the clerical condemnation and its victim's reaction are expressed not only dramatically, but also in numerical symbolism and a musical quotation. As the audience learns from outraged members of Linz's Lutheran congregation, Pastor Hizler, having been informed by contacts in Württemberg that Kepler once confessed his problems with the concept of the literal rather than spiritual presence of Christ's body and blood in the Eucharist, has insisted that Kepler sign a formal confession stating his religious adaptation. Kepler has refused to do so, not wishing to deceive himself or the Church. As a consequence, Pastor Hizler has publicly banned Kepler from communion and even declared publicly that the scientist is no longer welcome in his church. The pastor defends his action by pointing to directions received from the church consistory, which disallow that lay persons interpret articles of the faith according to their individual conscience. (He willfully ignores that, as he well knows, Kepler completed studies in theology before accepting appointment as a mathematician and astronomer, and can therefore hardly be considered a mere layman.)

All through this duet, the musical symbolism is eloquent. As already mentioned in a previous chapter (in the context of SEVEN as an unharmonious number in this work), Pastor Hizler rejects Kepler's request that he be allowed to follow his own conscience in 7 phrases. Each begins with the same 4 measures—a 2-bar component and its repetition. Lulled by the noticeable repetition, listeners risk missing the tonal sequence that is concealed in the fragment straddling the boundary: a transposition of Bach's famous signature, B-A-C-H.

EXAMPLE 14: Hizler attacking Kepler's "heretical" belief [11 C-I]

(I did what the consistory ordered. He interprets the faith as he chooses and despises the doctrines of the Church.)

The uppermost bracket in the example shows A♭–G–B♭–A♮, a transposition of B♭–A–C–B♮. The veiled allusion to Bach in the form of this indirect representation of his tonal signature is likely to pass unnoticed until Kepler isolates the tonal sequence in his replies. To the words that protest Hizler's denigration of his sincere self-questioning, Kepler even completes the line in a way that comes very close to the full form Bach used for the third fugal subject in the crowning piece of *The Art of the Fugue*:

EXAMPLE 15: Kepler defending his conscience, in Bachian humility [11 D5-6]

third fugal subject in "Fuga a 3 sogetti," *The Art of the Fugue* [8]

On three occasions does the vocal phrasing of Kepler's interjections cut across the structural boundary in Hizler's seven identical phrase beginnings so as to evoke "B-A-C-H."[9] With this tonal nod to the pious composer who was to live a century after Kepler, Hindemith highlights words that, though separated by many measures, form a sequence: Kepler's incredulous objection ("As though my belief were heresy") and his pain over the implacable wording used in the Lutherans' Augsburg Confession ("An explanation, unconsidered, careless") lead to the affirmation of his faith regarding Christ's presence ("As a spiritual effect alone").

In music interpreting Christian concepts, the number 3 is traditionally used as a symbol of the Trinity. When juxtaposed with 4 (the number conventionally associated with the elements of earthly life: earth, water, air, and fire), 3 also stands for the spiritual in contrast to the material. While Kepler's 3 interjections, focusing on an aspect of the musical line without destroying the whole, are spiritual, the pastor himself is more brutal: 4 times does he himself interrupt the identical component launching his phrases in order to affirm his *idée fixe*. The musical material in which he expresses these interruptions is heard 7 times, and all 7 cases underscore the word "Gegenwart" [presence], the concept with whose literalness the pastor is so singularly obsessed. In each case, the vocal line describes a similar melodic contour in Hizler's highest range, the strings fall abruptly silent, and the brass family, elsewhere in this duet locked in a unison on A, spreads to a sudden 7-part backdrop: a 6-pitch chord propped by a soft rumbling in drum and cymbal.

[8]As many Bach scholars agree, the primary fugal subject in *The Art of the Fugue* with its initial perfect fifth stands for the divine, while the semitone-studded third subject in the final fugue embodies flawed and sinful human nature in general, and this humble composer's repentant genuflection in front of his Maker in particular. (One is reminded of Hindemith's admiration for Bach as expressed a few years earlier in *Johann Sebastian Bach, heritage and obligation* [New Haven: Yale University Press, 1952].)

[9]See at [11 D4], [11 D12], and [11 G6].

The "unharmonious 7 " constituted of 4 + 3—or, material + spiritual components—also plays a role in the second focal scene of act II, where Hindemith explores another aspect of Kepler's conscientious attitude toward the world and his professional duty. The meeting between Kepler and Susanna in which they agree to join their lives is musically cast in two ternary structures: recitative–Lied–recitative (at [14]) and arioso–elegy–arioso (at [15]). The first shows the lingering impact of Pastor Hizler's antagonism: the initial lines in Susanna's opening recitative, in which she ponders the courage she mustered to defend Kepler against the dictatorial pastor, are interrupted 7 times by woodwind figures, echoing the 7 interruptions with Pastor Hizler's insistence on belief in "real presence."

The Spiritual Message in Hindemith's Quotations

Hindemith's opera integrates extant musical material of three different kinds; each conveys a message. With his skillfully crafted allusion to Bach's musical signature, the composer likens his protagonist to another extraordinary man with whom, he seems to suggest, Kepler had much in common. As his late masterpiece shows, Bach regarded professional activity, much as it was ostensibly dictated by the requirements of his employment and geared to practical ends, as a way to serve God. Filled with the humility that he found so aptly expressed in the semitonal writhing of his name's musical "lettering," Bach strove to make his life's work a single contribution to the greater glory of God. The same, Hindemith asserts with the repeated quotation of the BACH motif in the condemnation scene, holds true for Kepler.

The second kind of musical quotation corresponds more narrowly with what the word usually connotes: a recognizable entity, known from another context, is inserted into a new environment, from whose musical language it differs. Its hermeneutic background opens a crucial window of interpretation. Where other composers prefer motifs or themes, Hindemith likes to quote entire songs.[10] Of the three quotations, Hindemith himself identifies

[10]In his earlier opera, *Mathis der Maler*, he quotes an early hymn, "Es sungen drei Engel ein süßen Gesang," as well as a 16th-century folk song, a version of the "Evangelisches Jubellied" from Mathis's (Grünewald's) lifetime, and the Corpus Christi sequence "Lauda Sion Salvatorem." In his ballet music *Nobilissima Visione*, inspired by a fresco cycle on the life of St. Francis of Assisi which Giotto painted in the Bardi chapel of the Florentine church of Santa Croce, Hindemith embodies the dancing protagonist in the musical emblem of a trouvère song, a hymn to the month of May composed at the time of the Tuscan saint. For more on the latter, see the chapter on this work in S. Bruhn, *Musical Ekphrasis: Composers Responding to Poetry and Painting* (Hillsdale, NY: Pendragon Press, 2000), especially pp. 457-459.

one—probably above all because he wishes to give credit for the poem, which is Kepler's own. A footnote in the score alerts musicians (and, via the customary reprint in program books, audiences) to the source of the tune to which Kepler's lament in act I is set: a hymn that Johann Hermann Schein (1586-1630), a contemporary of Kepler and one of the leading composers to develop early Lutheran church music, wrote in 1627.[11] The composer's name, which can simply translate as "shine" but also evokes "semblance" or "sham," may have recommended him to Hindemith as a fitting source for a tune to a text about the insubstantiality of the human life.

Hindemith adapts Schein's melody to Kepler's poem by adding a varied repetition as phrase 3 (cf. Exx. 16 and 17). At [27 C] in act V, when Kepler acknowledges anew that ultimate harmony cannot be gained in life but awaits humans only after death, the entire tune recurs in a new 4-part harmonization by the strings and woodwinds, with Kepler singing the central phrase to the words: "Dying is necessary to bring it about."

EXAMPLE 16: Johann Hermann Schein, "Seligkeit, Fried, Freud und Ruh"

EXAMPLE 17: Kepler's song of mourning (from stanza 2, literal translation)

[11]"Seligkeit, Fried, Freud und Ruh" [Bliss, peace, joy, and tranquility], J.H. Schein, *Cantional* [*Neue Ausgabe sämtlicher Werke* II/2] (Kassel: Bärenreiter, 1967), p. 81. More on these verses in the context of Hindemith's integration of Kepler's poetry, in part 7 below.

Later in the scene, after the Regensburg pastor's visit, Kepler regrets that even the proximity of life's end does not grant relief from discord; only the body's death can free the soul and thus allow it to partake of harmony. As he muses about this insight, expressions of hope alternate with the four lines of the epitaph the historical Kepler composed for his tombstone. These inserts, verbally highlighted through the use of Latin, are also sung to the phrases of the Schein tune, a choice that seems fitting given their pious acceptance of death.

EXAMPLE 18: The Schein tune underlying the tombstone epitaph

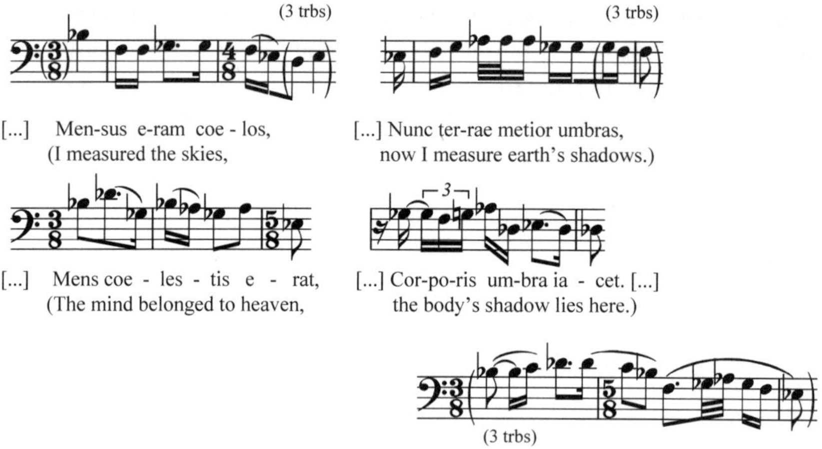

Hindemith is less forthcoming regarding the provenance of another hymn, and scholars have correspondingly neglected to identify its source, although the implications of this knowledge would seem significant. In act II, just before the Lutheran congregation at Linz emerges from the service during which their pastor has banned Kepler from communion, they are heard singing "Dank sagen wir alle Gott, unserm Herrn Christo" [We all give thanks to God, our Lord Christ]. The tune they intone is a close variant of one appearing to the same words in Schein's *Cantional*.[12] The shared source for this communal hymn and Kepler's private lament seems highly significant: the music establishes that Kepler's religious attitude was of a piece with that of the Lutheran community of which he considered himself a faithful member, and by no means rebellious, as its clergy chose to claim.

The doctrinal dogmatism whose Lutheran variant overshadowed Kepler's personal life led to the long series of battles between Catholics and Reformed Christians that would later become known as the Thirty Years

[12]"Grates nunc omnes" or "Gebetlein", *Cantional*, part 1, p. 29.

War. Hindemith captures its spirit in a musical quotation, and its problems in the *contrafactum* he wrote to match it. The tune of the "Old War Song" in act V, with which the electors urge and finally compel the emperor to dismiss Wallenstein, is taken from a song listed in Böhme's *Altdeutsches Liederbuch* under the title "Kriegsfreud und Kriegsleid" [War's joy and sorrow].[13] The earliest known prints link this song to the Thirty Years War.[14] Despite the ostensible balance of the positive and negative aspects of armed strife in the song's title, the twelve stanzas clearly glorify war. Stanzas 3-4 briefly touch on the sorrow of sweethearts left behind, only to advocate a heroic and dashing farewell; stanzas 10-11 speak equally briefly of lives lost and comrades buried, only to argue that since we all have to die, death on the battle field, "fresh and free, risking one's blood for honor and glory," is by far preferable to any other form of death.

Conversely, the words Hindemith has the electors exchange with their emperor to this tune in *Die Harmonie der Welt* have nothing of soldiers' smugness. Their complaint that the war is destroying the empire, that no-one can breathe freely, and that neither army nor royal leaders have anything but their own power in mind, comes close to a condemnation of warfare in general; the emperor's justification, that without Wallenstein's help the Catholic faith might not survive against the threat of the Reformation, sounds unconvincing and is correspondingly ill-received.

The spirit of this *contrafactum* is also enhanced by musical means. Matching the assertive tune, the song is composed in a $\frac{4}{4}$ time that is to sound "in the tempo of a heavy march," with the melodic lines bolstered by homophonic winds and percussion. The simple time is, however, repeatedly called into question. In some cases, strings and timpani juxtapose a different order (irregularly spaced snippets in a waltzy $\frac{3}{4}$ time in the theme and variation 3, figures in combinations of $\frac{5}{4}$ and $\frac{6}{4}$ time in variation 1). In other cases, contractions in the tune itself lead to the sudden truncation of some measures and cause a metaphoric stumble in the soldiers' heavy-booted see-saw. This happens notably in variations 2 and 4, to the emperor's attempts at justifying the necessity of war as a means toward preserving Catholicism and the Habsburg monarchy. With his musical signals of disorder and mismatched proportions, Hindemith passes an unambiguous judgment about this war—and with it, presumably, about the general idea of forcefully imposing a single interpretation of a facet within a shared faith.

[13]Franz Magnus Böhme, *Altdeutsches Liederbuch: Volkslieder der Deutschen nach Wort und Weise aus dem 12. bis zum 17. Jahrhundert* (Hildesheim: Olms, 1966), pp. 529-531, no. 424.

[14]With the complete text, the song is first documented in a 1646 book; an incomplete version on a loose sheet issued in 1622 suggests that it may have been sung earlier.

The third kind of musical "quotation" of sorts is a timbral allusion. In the context of the epitaph, Hindemith makes use of the distinctive sound of three trombones playing unison in *pp*. When the protagonist, approaching death, quotes the Schein tune to which he had earlier sung his lament, trombones supplement the tones that his words omit (see the parentheses in Ex. 18 above) and anticipate the two central phrases in stretto. Nowhere else in the opera are the trombones employed in just this way.

However, a notable predecessor of the sound of three trombones playing in very soft unison exists in the overture to *Mathis der Maler*. In the German tradition, trombones have a long history of suggesting divine judgment and punishment. In concert music, famous trombone statements with this connotation occur in Mozart's Requiem, in the last movement of Beethoven's Ninth Symphony, in Berlioz's *Grande messe des morts*, etc. In sacred music, trombones were often deployed as a latter-day emblem of the angelic instruments announcing divine punishment, which Luther, in his translation of the Bible, had rendered as trombones.[15] In opera, too, the sound of trombones is featured as a timbral signifier of divine pronouncements; see, e.g., the oracle scene of Gluck's *Alceste,* the "sacrifice" scene in Mozart's *Idomeneo*, the judgment-at-dinner scene in *Don Giovanni*. Because of the Christian concept of the Trinity, the musical rendition of divine judgment was often expressed by *three* trombones playing as *one*, in unison. As I have argued elsewhere,[16] Hindemith's choice of scoring stanza 1 of the chorale "Es sungen drei Engel ein süßen Gesang" in his *Mathis* overture with three trombones playing *pp* in unison announces, in a musically couched way, that the opera is about divine judgment, about the question whether the protagonist will attain vindication for the major decisions he made in his life.

The use of the same timbral signifier in this opera clearly bears the same connotation. Sounding immediately before Kepler's death (and only there), interspersed with his final musing on whether harmony can be achieved on earth, the trombones seem to emphasize—comfortingly—that what Kepler will be judged by is his (professional and private) metaphysical quest.

[15]English Bible translations have trumpets (see, e.g., Revelation 8:2 etc.); the Latin Vulgate speaks of tubas. A composer's timbral symbolism presumably reflects the cultural tradition with which he or she grew up. In the Hebrew context, the instrument alluded to was, of course, none of the instrument of the modern orchestra but the ram's horn or *shofar*.

[16]Siglind Bruhn, *The Temptation of Paul Hindemith:* Mathis der Maler *as a Spiritual Testimony* (Stuyvesant, NY: Pendragon Press, 1998), pp. 290-291.

The *I Ching* in the Life of Joseph Knecht

Metaphysics in the *Glass Bead Game* is of a very different kind. Next to music and meditation, the third force pervading every level of Hesse's work is the wisdom of ancient China. Introduced as a motif in movement I of the discursive glass bead game and endorsed with an extensive quotation from a book of annals compiled ca. 2500 years before the protagonist's life time, the Chinese view of music as an ethical force and barometer informs the World of the Mind as a whole. China is among the cultures to which Castalia bows when naming the ancestors of its intellectual-artistic game. Chinese philologists issued the "Warning Cry" urging the development of an international and transdisciplinary language of symbols that, "like the ancient Chinese script, should be able to express the most complex matters graphically, without excluding individual imagination and inventiveness" (p. 37). The chronicler compares the wealth of the game language with the Chinese script, remarking that by the year 2400, the game's officially accepted symbols long "exceeded the number of the ancient Chinese ideographs" (p. 42).

In each of the three games for which details are given, the great texts of the ancient Asian cultures feature as thematic components. The narrator refers to one of India's ancient scriptures when he substantiates his abstract explanations by specifying that a game "might start from a given astronomical configuration, or from the actual theme of a Bach fugue, or from a sentence out of Leibniz or the Upanishads ..." (p. 40). The game Joseph Knecht and his friend Fritz Tegularius sketch as students in an advanced course on glass bead game composition "began with a rhythmic analysis of a fugal theme and in the center of it was a sentence attributed to Confucius" (p. 120). And Knecht's first ceremonial game, as discussed above, is famously based "on the ancient ritual Confucian pattern for the building of a Chinese house" (p. 245).

Beyond this thematic integration of sentences from the Eastern wisdom traditions in actual game compositions, the presence of the polar forces *yin* and *yang* in all aspects of mind and matter—the basic tenet of both Daoist and Confucian scriptures—plays a significant role. This concept dates back to the earliest centuries of the Chinese civilization. Confucius (*Lun Yü* IX: 16) urges his disciples not to ponder the river but instead the great law that causes the water to flow, its depth, its speed to change, and even its bed to move over time; Laozi calls the Dao the unchangeable law of change. Both believed that all that exists began with the primeval complementarity, represented in the well-known symbol whose bright *yang* contains a kernel of dark *yin* and vice versa.

Five times in the course of the story does Joseph Knecht explicitly refer to the *yin-yang* dyad. In the introduction to *The Glass Bead Game*, the narrator quotes from the *Annals of Lü Buwei* the beginning of a poem that makes music the primeval ancestor of the polar forces; in the wording of the Knoblock–Riegel translation: “The origins of music / Lie in the distant past: / Born of measurement. / Founded by Grand One, / Grand One brought forth the Dyadic Couple; / The Dyadic Couple brought forth Yin and Yang.”[17]

During one of the courses in glass bead game composition that Knecht attends, he is suddenly struck by the awareness that in the spirit of the game, “everything was actually all-meaningful, that every symbol and combination of symbols led not hither and yon [...] but into the center, the mystery and innermost heart of the world, into primal knowledge. [...] every classical or artistic formulation was, I realized in that flashing moment, if seen with a truly meditative mind, nothing but a direct route into the interior of the cosmic mystery, where in the alternation between inhaling and exhaling, between heaven and earth, between Yin and Yang, holiness is forever being created” (pp. 118-119).

Upon hearing first rumors that he might be elected as successor to the recently deceased *magister ludi*, Knecht has a vision. A mentor and his young student follow one another as if drawn along the wires of some mechanism. The image is that of a kind of transtemporal round dance in which he, Joseph, is alternately the old man and the boy, “now revering and now revered, now leading, now obeying”—forever changing, forever the same. Knecht interprets his vision with reference to the great polarity of all things: “there evolved [...] the insight that this meaningful and meaningless cycle of master and pupil, this courtship of wisdom by youth, of youth by wisdom, this endless, oscillating game was the symbol of Castalia. In fact it was the game of life in general, divided into old and young, day and night, yang and yin” (p. 221).

Before telling of his protagonist’s startling defection from Castalia, the chronicler provides an extensive retrospective, reflecting on Knecht’s worry about the relationship between the World of the Mind and the world-at-large as well as on his concerned sympathy for Tegularius as a portent of what Castalians might become. The summary remark is again phrased in Chinese terms: “The two tendencies or antipodes of this life, its Yin and Yang, were the conservative tendency toward loyalty, toward unstinting service of the hierarchy on the one hand, and on the other hand the tendency toward ‘awakening,’ toward advancing, toward apprehending reality” (p. 274).

[17]For the larger context of these lines, which appear as prose in Hesse’s work (p. 29), see the very beginning of the chapters given in my Appendix .

Finally, at the outset of his decisive conversation with Plinio Designori, musing whether it is correct that "every human being on earth can fundamentally hold a dialogue with every other" or whether, conversely, "there are no two persons in the world between whom genuine, whole, intimate understanding is possible," Knecht judges that "the one statement is as true as the other. It is Yin and Yang, day and night; both are right" (p. 294).

A third step toward ever greater specificity in the metaphysical system leads beyond the Eastern wisdom traditions in general and the basic complementarity of *yin* and *yang* in ancient Chinese thinking in particular to the oldest compilation of Chinese wisdom, the *I Ching*. Before I explore the three explicit references to hexagrams and the judgments attached to them, I would like to recall an event from Hesse's own life that seems germane here. Hesse did not only admire and write about, he actually practiced the *I Ching*. In the spring of 1954, Theodor Heuss, Germany's first president after World War II, invited Hesse to travel north to be received into the distinguished order *Pour le mérite*. Hesse, by then almost 77 years old, generally no longer accepted awards. But since Heuss was an old acquaintance, he felt torn between reluctance and obligation. As he wrote Heuss in a letter dated Pentecost 1954, he consulted the *I Ching* and, having interpreted the oracle's pronouncement, found that it was favorable for him at this time to accept invitation and honor; so he would go.

A satisfactory introduction to the *I Ching* must remain beyond the scope of this study. Suffice it here to recall only the fundamental data. As Hesse knew, who had read and reviewed Richard Wilhelm's commented German translation of the work, the *I Ching* allegedly began its existence as early as 24 centuries B.C.E., when the initial trigrams, and the idea of joining them to hexagrams, was first conceived. In time, each hexagram acquired a name, a "judgment," a statement explaining the significance of each of the six lines in the context, and finally, a layer of commentaries.[18] Confucius was long credited with having compiled the *I Ching* and written most of its extensive commentary. Today's scholars believe that the genesis was quite a bit more complex, and that Confucius' own input constitutes only a portion.

[18]As Richard Wilhelm explains, Chinese literature credits four holy men with the authorship of the *Book of Changes*: Fu Hsi, King Wen, the Duke of Chou, and Confucius. Fu Hsi, a legendary cultural hero and sage from mythological times, is said to have invented the trigrams, King Wen of the Zhou (reigned 1171-1122) to have developed the hexagrams, the Duke of Zhou (died 1094) to have composed the judgments, and Confucius to have written (part of) the commentary. Both my brief summary and the subsequent comments on the hexagrams integrated into Hesse's work are based on *The I Ching or Book of Changes*, the Richard Wilhelm translation rendered into English by Cary F. Baynes; foreword by C.G. Jung (Princeton, NJ: Princeton University Press, 1950).

The *I Ching* explores the sixty-four hexagrams that can be built with two kinds of lines—the unbroken (or strong, solid, bright, male, active) *yang* line and the broken (or weak, yielding, shadowy, female, submissive) *yin* line. Which of the two falls on each of the six places in the hexagram is determined in an elaborate counting process undertaken originally with 50 yarrow sticks, later also with coins. The six stacked lines, read from bottom to top, are interpreted in several ways, the most common of which are: in terms of the image created by the two superimposed trigrams (lines 1-3 below 4-6); in terms of the particular impact of individual lines, which relies on the aspect a *yang* or *yin* stresses depending on the relative strength of its place; in terms of the added meaning contributed by the two nuclear trigrams (lines 2-4 and 3-5); and occasionally in terms of the three line-pairs (1-2, referring to earth; 3-4, referring to the human realm; and 5-6, referring to heaven.) In all this, *yin* and *yang* with their multifaceted significations represent not opposites but counterparts or complements. For a Western mind, this is easier to grasp in some attributes than in others. The pair light–shadow is helpful; the interpretations of *yin–yang* as matter–spirit or as nature–mind (Castalia's challenge) also convey a sense of the mutual dependence.

The sense in which a hexagram is understood to be relevant to a given situation relies on the assumption that what the complex counting process yields in a certain moment reflects the unique quality of that moment. Chinese thinkers believe causality to be but a marginal force compared to chance, which is pervasive. The emphasis is rarely on things as they are or will come to pass (the West's primary concern), but usually on the way in which conditions may change or be changed. The hexagrams are never read as indicators of factual circumstances but as pointers toward transformative tendencies. Correctly understood, each moment with its quality and propensity for change can be met with suitable or unsuitable action. As Wilhelm explains in his preface to the German translation of the *I Ching*, Western oracles typically inform their clients about what they can expect to happen, whereas the *I Ching* advises them on what they might do or refrain from doing so as to strengthen a particular moment's favorable quality and weaken its inauspicious aspects. To the Chinese understanding, humans are not the passive recipients but the active co-shapers of their destiny.

To get a sense of the eight trigrams is relatively easy if one associates each of them in three ways: with an element, a family position, and various attributes. The interpretation in terms of family positions works wonders as a memory aid once one accepts that sons are understood to be more closely related to their mother while daughters share more with their father. The reading in terms of eight elements and their qualities or attributes provides a first step into the mystery of hexagram interpretation.

The two basic trigrams consist of three identical lines each:

☰ triple *yang* stands for heaven, the creative principle, father, and strong;

☷ three *yin* lines signify earth, receptivity, mother, and yielding.

When the mother's favorites—the sons—are born, their place in the birth order is marked by the ascending position of their single *yang* line:

☳ Strong *yang* below duple yielding *yin* is ethereal energy (as, e.g., in thunderstorms), an arousing and inciting force, the first son.

☵ Strong *yang* in the center between yielding forces is water in motion (in flowing rivers, sprouting wells, plunging waterfalls, rising vapor, and falling rain); it is the dangerous or abysmal force, the second son.

☶ Strong *yang* atop two layers of yielding *yin* is the stabilizing principle, the mountain, calm and static, the third son.

The father's favorites—three daughters—define their attributes similarly:

☴ Yielding *yin* below duple *yang* results in a gentle but penetrating force; it is wind or wood, and first daughter.

☲ Yielding *yin* surrounded by *yang* (second daughter) equals fire, to the Chinese an element that never exists alone but attaches itself to matter and feeds on oxygen; it is clinging but also bright and light-giving.

☱ Yielding *yin* above duple strong *yang* creates serenity and joyfulness; this is the lake or other quiet water, and the third daughter.

Joseph Knecht comes into contact with the *I Ching* at three prominent junctures in his development. During his years of free study, when he sets himself the task of tracing the intellectual background of a glass bead game he once played, he studies Chinese language and literature and then requests of the educational authority that he be allowed to take an introductory course in the *I Ching*. Since no such course is being offered in Castalia's colleges—any teaching with an irrational component is at odds with the Confucian spirit reigning in this era—Knecht is sent to "Elder Brother," the Sinologist turned Daoist recluse, whom he visits at his bamboo grove. This self-styled Chinese hermit casts the yarrow sticks in order to determine whether the time is favorable for granting the requested instruction, and whether the young visitor is truly serious and submissive and should therefore be allowed to stay. As the oracle is encouraging, Knecht may remain for several months, receiving training in the spirit and practice of the *Book of Changes* and a thorough introduction to Daoist thinking.

A few years later, when Knecht is dispatched as a *pro forma* glass bead game teacher and *de facto* diplomat to a monastery, he himself consults the *I Ching* by means of the yarrow-stalk ritual before he sets out. Among the Benedictines he encounters an abbot enthusiastic about the *Book of Changes*.

Their shared admiration for the Chinese wisdom tradition and Knecht's ability to foster the abbot's understanding become an important element in the success of Knecht's mission.

When the newly appointed *magister ludi* plans his first official game, he remembers a commentary on the *I Ching* about traditional Chinese house building that had once delighted him with its combination of mythic order and speculative scholarliness. Having decided to design a ceremonial performance based on this structure, he writes to Elder Brother in the hope that the Daoist hermit with his intimate knowledge of ancient Chinese thinking will consent to assist him in his preparations. The invitation is rejected—in three phrases taken from various judgments contained in the *I Ching*.

So what are these hexagrams and their judgments? I want to begin with the oracle Knecht receives on the eve of his departure for the monastery. The yarrow-stalk ritual produces hexagram no. 56, *Lü* or "The Wanderer." The ideogram consists of a mountain with fire on its top. The judgment stresses the idea of the power of smallness: ䷷

> The Wanderer. Success through smallness.
> Perseverance brings good fortune to the wanderer.[19]

Richard Wilhelm comments:

> When a man is a wanderer and stranger, he should not be gruff or overbearing. He has no large circle of acquaintances, therefore he should not give himself airs. He must be cautious and reserved; in this way he protects himself from evil. If he is obliging toward others, he wins success.

Hesse has Knecht's attention focus on one of the line statements. For the *yin* line on the second place in this particular hexagram, the *I Ching* has the following comment:

> The wanderer comes to the inn.
> He has his property with him.
> He wins the perseverance of a young servant.

On the surface, this observation seems above all convenient for narrative credibility: Hesse found it easy to have Knecht, during his period among the Benedictines, encounter a young person who, attracted like so many others by his luminous and serene personality, approaches him with "perseverance"

[19]The hexagram numbers refer to the *I Ching* version on which Wilhelm based his book. I give Baynes's wording, which differs slightly from that of the Winston's translation of Hesse's quotations from Wilhelm's German (pp. 151-152 in *The Glass Bead Game*).

and attempts to be of service. Yet the curious way in which the chronicler de-emphasizes the young man's importance, calling him "the young monk, concerning whose life we have no further information" (p. 187) and stressing that "this young man subsequently played no part in Joseph Knecht's life" (p. 160), suggests that the person appears in the narrative to justify the *I Ching* comment rather than the other way around. Should this speculation be correct, it would lead initiates of the *Book of Changes* to the sign resulting when the line thus singled out—the second from the bottom—converges from *yin* to *yang*. When the fire is no longer on top of a mountain but instead above wood and gentle wind, the meaning changes from "wanderer" to "cauldron" (no. 50), one of the most positive hexagrams in the entire *I Ching*. The Chinese word associated with the sign suggests a combination of secular and sacred functions: a *ding* or cauldron holds food for family and guests but is also used in ancestor worship. Hence the person faced with the augury "Lü" may be destined to be a wanderer and respectful guest in the immediate future; if he is flexible, he has the potential to become a vessel of tangible as well as spiritual sustenance.

䷷ → ䷱

In the context of Knecht's request for instruction in the *I Ching*, Elder Brother's yarrow-stalk ritual results in the sign no. 4, *Mêng*. This ideogram, which shows moving water beneath a mountain (the volatile force at the foot of the stable principle), bears the name "Youthful Folly." The judgment is:

䷃

> Youthful folly has success
> It is not I who seek the young fool;
> The young fool seeks me.
> At the first oracle I inform him.
> If he asks two or three times, it is importunity.
> If he importunes, I give him no information.
> Perseverance furthers.

Wilhelm's explanation reads in part:

> In the time of youth, folly is not an evil. One may succeed in spite of it, provided one finds an experienced teacher and has the right attitude toward him. This means, first of all, that the youth himself must be conscious of his lack of experience and must seek out the teacher. Without this modesty and this interest there is no guarantee that he has the necessary receptivity, which should express itself in respectful acceptance of the teacher. This is the reason why the teacher must wait to be sought out instead of offering himself. Only thus can the instruction take place at the right time and in the right way.

In the course of his training at the bamboo grove, Knecht attempts repeatedly to steer the conversation toward the glass bead game. He is still uncertain whether the game should be his future life's focus—whether it is truly "the highest and most demanding master," whom he has vowed to serve—and longs to hear the Daoist's opinion. Yet any responses he receives are phrased in pronouncements from the *I Ching*. The chronicler (p. 132) quotes two such replies: "Dense clouds, no rain" and "Nobility is without flaw" (literally, in Hesse's German as in the *I Ching*, "the noble man is without flaw/blame.")

Tracing the origin of these two sentences proves rewarding. The first stems from the judgment for hexagram no. 9, "The Taming Power of the

䷈

Small," which speaks of the effect that may be caused by things that seem insignificant and meek. The image shows the gently penetrating wind above the creative force of heaven. In this constellation, even the generative power of five strong *yang* lines is temporarily checked by the one *yin* line: the dense clouds, ready to bless the land with their nourishing water, are kept flexible and thus temporarily held in check.

䷓

The second of Elder Brother's cryptic replies is from hexagram no. 20, "Contemplation." The words are not, however, part of the judgment, but occur in the comment on the fifth line, which Baynes translates from Wilhelm as: "Contemplation of my life. The superior man is without blame." Once again, the focus on a single line may remind those seeking wisdom of the ubiquitous potential for change that is the basic truth of all life. Were the line on which the seeker's eye happened to fall—

䷓ → ䷖

here, the second from the top—to change from strong *yang* to yielding *yin*, the result would be hexagram 23, "Splitting Apart." This is precisely what the Daoist hermit seeks to avoid for himself and against what he warns his apprentice. Thinking about the glass bead game before the newly opened horizon of the *I Ching* may appear to Knecht as merely a way of pondering his life—and as such without flaw. But it risks breaking his concentration, causing distraction, and leading to disharmony.

The phrases with which Elder Brother rejects the invitation issued by the man who was once his pupil and is now the youngest of Castalia's dignified masters, serves superficially to convey the message that he is unwilling to leave his bamboo hermitage. But there is more in the three lines that precede the formal Chinese greeting (p. 247). The opening, "It would be honorable to behold the great man," appears repeatedly in the *I Ching*, namely in the judgments of nos. 6, 39, 45, and 57 and in line statements of hexagrams 1 and 39. The second phrase, "But movement leads to obstacles," occurs in four of the line statement for hexagram 39 and thus seems to confirm this

ideogram as the intended source. What is more, the statement for the uppermost line contains both sentences, in the exact wording found in Hesse's original. Wilhelm explains that the context suggests a man who, having developed to realms beyond the world and its bustle, might be tempted to stay aloof and simply ignore its problems. But this is not an auspicious path: he should not strive for his private salvation. Instead, his experience and unusual inner freedom are precisely the attributes that compel him to get involve and effect great and beneficial changes.

At first glance, the two sentences appear as a comment on the Daoist hermit's excuse for not obliging Knecht's request for assistance. But on closer inspection, one wonders whether the person who should resist the temptation to stay aloof and ignore the world's problems is not rather the *magister ludi*. Contrary to Elder Brother, who has segregated himself not only from the world, but even from Castalia, Joseph Knecht has long been aware of the bustle outside the borders of the protected province of the mind, and has begun to perceive his order's aloofness as a problem. If the line that carries this combined meaning were to change—if the *yin* at the top of hexagram no. 39, "Obstruction," were to transform itself to *yang*—the result would be hexagram no. 53, "Development – Gradual Progress." Read in the spirit of the *I Ching*, Elder Brother's reply thus expresses not only a recluse's unwillingness to interrupt his solitude, but a veiled recommendation for the note's recipient to continue his gradual progress toward overcoming the obstacles that keep him unduly aloof from the concerns of the larger world.

The third sentence in the hermit's reply, "Let two small bowls be used for the sacrifice," seems to contain a different kind of warning. Hesse's German wording is a verbatim quotation from the judgment for hexagram no. 41, "Decrease." Beginning once again at the surface, the seeker may believe this source to confirm that for Elder Brother, "beholding the great man" might not be an unequivocal gain but rather a decrease, i.e., a distraction from the pursuit of his spiritual goals. However, as Wilhelm explains this sign,

> Decrease does not under all circumstances mean something bad. Increase and decrease come in their own time. [...] If the time of scanty resources brings out an inner truth, one must not feel ashamed of simplicity. For simplicity is then the very thing needed to provide inner strength for further undertakings.

Applied to Joseph Knecht, this reminder to risk a decrease in station and explore simplicity seems like a prophesy of the decision he will reach eight years later, to give up all his privileges and seek direct impact on the world.

So much for the narrative significance of the three situations in which Knecht comes into contact with the *I Ching*. Ultimately more important for an understanding of Hesse's intention is the role the various judgments play with regard to the overall message of *The Glass Bead Game*. Five topics can be filtered out from the various wisdom pronouncements discussed above:

(1) Learning is crucial, but favorable only if the seeker recognizes his ignorance ("youthful folly"), seeks a mentor, approaches him respectfully, and queries him sparingly.

(2) Smallness has its own success and power. Even a wanderer—a person far from his rightful home—may have significant impact if he is well equipped ("has his property with him"). An apparent "decrease" may lead to actual gain.

(3) A glass bead game master is a great man and honorable to behold; yet he should seek to become a sacred vessel that will contain what meets material and spiritual needs.

(4) Aloofness is a blemish; no degree of individual progress is an excuse for feeling superior to the larger world.

(5) Being split apart, in one's own soul or in a community, results in disharmony. The ultimate goal is the higher unity between mind and nature, to be achieved both in contemplation and in action.

Each of these insights can be read as relevant to the overall themes of Hesse's work. Castalia with its glass bead game has begun to stagnate when its hierarchy no longer recognized the need for further insights and developments. The Castalian Music Master, once lovingly referred to as "the great would-be small," is portrayed as a living model of the truth that the ostensibly smallest may be the greatest, a state the *magister ludi* also approaches when he begins reminding people that the etymology of his title derives from "schoolmaster." Castalia as a whole would also do well to remember its position as a guest on the larger world's grounds, and substitute conceit and class arrogance with approachability and concern for its hosts. Splendid ceremonies cannot replace real service to the world's needs for enlightenment and serenity. Instead of joining forces with all that the realm of "real life" has to offer, Castalia's elitism deepens the chasm between the World of the Mind and the world-at-large. As a result, its own values are withering over time until its goal—ever more encompassing synthesis and harmony—vanishes out of sight beyond rifts that can no longer be healed.

Part VI
COSMIC HARMONY AND HUMAN EXPERIENCE

PLATE 6: Image of the November 1577 comet, which Kepler witnessed as a child, as captured in a contemporary etching

Cosmic Events and Their Impact on Earth

Humans are supremely susceptible to any harmonious proportions in the properties of their natural environment. The Greeks famously believed that a person's emotions and passions reflect changes of the musical scale; Plato (*Republic*, Book 3) held that Dorian melodies are nobler than Phrygian ones, that the Lydian mode is expressive of sorrow, etc. The conviction was that impressions received through the ear had a direct impact on the "soul"—that indescribable entity that ranges from visceral reactions all the way to states of spiritual rapture. To varying degrees, the same susceptibility applies in the case of harmony in the combinations of colors, enclosing spaces, movements, and other aspects of what constitutes a person's experiential world.

When, still in the *Republic*, Plato describes the tyrant as a man who is 729 times as bad as a good man, readers are initially puzzled. Is this strange number just a way of saying "many," a case of literary license? To thinkers trained in the numbers corresponding with the various degrees of harmony, however, the clue is clear: 729 is 3 x 3 x 3 x 3 x 3 x 3. Since the number 3—as the third partial or multiple—defines the upper and lower fifths of a fundamental tone, the sixth fifth in both directions corresponds to the tritone, the worst possible dissonance in the musical system of ancient Greece and, in fact, in all of Western culture before the 20th century. "What Plato evaluated by the number 729 was the relation between the good man and the tyrant as that of the greatest possible tension within a civilized system."[1]

If discomforts of body, soul, or mind are expressions of dissonance, exposure to consonances is bound to act as a healing assistant, if not a remedy. Controlled studies of the influence of music and art therapy confirm this link between external and internal harmony. What responds are the senses; one's analytical faculties can only calculate and then corroborate what is already happening beyond their control. For a person to feel uncomfortable or at ease with certain tonal relationships, certain ratios in the world of shapes, or certain coincidences of the color spectrum, no intellectual awareness of reigning conditions, let alone of their causes, is required. As Kepler writes about the disposition of the soul: "To find the appropriate proportion in sensible things

[1]Siegmund Levarie in Ernest McClain, *The Myth of Invariance: The Origin of the Gods, Mathematics and Music from the* Rg Veda *to Plato* (New York: Nicolas Hays, 1976), p. xii.

is to uncover and recognize and bring to light a similarity of that proportion in sensible things to some particular archetype of the truest harmony which is within the soul."[2]

Analogous truths pertain to larger bodies, such as societies. In a "political digression" added to Book III of his *Harmony of the World*, Kepler expressed this facetiously when he proposed an allegorical translation into political systems of the three ways a distance can be bisected (or a collection of numbers averaged)—the arithmetic, geometric, and harmonic means. Democracy, in which all members are equal, corresponds to the arithmetic mean, he jokes; aristocracy, in which a minority enjoys blatant feudal prerogatives, can be compared to the geometric mean; and monarchy, as the diplomatic combination of both, resembles the harmonic mean.

Among the most awe-inspiring impressions our senses receive are those pertaining to the heavenly bodies; it can be no coincidence that so many early cultures were convinced that the sun and the moon, the stars and the planets were divine beings with a powerful influence upon the events on earth. Enlightened members of the modern world believe to know that those bodies are neither divine nor otherwise alive and cannot, therefore, have any impact on our home planet other than the dangers caused by an occasional meteorite. But since Kepler was obliged to spend a fair portion of his time addressing questions of astrology, the study of the link between events in the sky and events on earth, it is useful to look a little closer at what this "soft science" signified for him.

Encyclopedias describe astrology variously as a system based on the belief that events on earth "are represented by," "are influenced by," or "reflect" the positions and movements of luminous entities in the sky. In each of these definitions, astrology is seen as aiming to elucidate how the motions of the heavenly bodies and their relative position to one another as seen from earth influence the earth's atmosphere as well as bodies, souls, and societies. The difference in wording, however, is essential. It leads directly to Kepler's reproaches against and qualified defense of the discipline.

One of Kepler's tasks as a professor of mathematics in Graz and later as the imperial mathematician at the court in Prague was to publish an annual calendar with prognoses for the coming year. It was expected that he calculate"what was going to happen" in terms of war and peace, droughts and deluges, as well as the fate of the country's most prominent citizens. Admiring readers, including many who dabbled in the field, hoped for reassurance that even undesirable events occurred according to some higher plan and, with adequate precautions, could ideally be minimized in their impact.

[2]Johannes Kepler, *The Harmony of the World*, Book IV, p. 295.

Kepler disappointed these expectations. He had thought long and hard about astrology and was eager to share his qualified viewpoint. In a nutshell: he loathed astrology's link to fortune-telling and astromantic practices, but was convinced that no susceptible being could fail to sense, and therefore respond to, the strong constellations that occur in the outermost sphere of their perceptible environment. In the preface to his treatise "On the More Certain Fundamentals of Astrology," he voices the pros and cons of his involvement with the field by acknowledging that

> there are those who say that it is not fitting for a serious and philosophical person to prostitute the reputation of one's talents and the honor of one's erudition in matters debased by the very trivial and very empty daily soothsayings; by furnishing this divining, one feeds the curiosity of the crowd and kindles the superstitions of weak dispositions as if with a tinder. [...] But if the desire to enlighten himself suffices as a person's motive (and such motives are approved for their own sake by wise men), then he cannot be stopped. [...] For even if a great number of precepts are taken up with the trifles [...], still those things that are in the hidden art of nature are not on that account trifles, nor should they be rejected as trifles. But gems must be gathered from the mire [...].[3]

This treatise originated as the first calendar Kepler prepared in his new position in Prague. Named imperial mathematician after the unexpected death of the respected Tycho Brahe, he still felt that his true vocation—had he only been in a financial situation to choose—would have been that of a philosopher. So he decided to state clearly what could be expected of him and of astrology, and what hopes should be buried as based on superstition. As a result, the explanatory segment of his calendar covers about two thirds of the text; the "prognoses" for the year 1602, the alleged purpose of the document, take up only one third. In 75 theses Kepler emphasizes factual observations, expounds some general principles: he explains the impact of direct or reflected light, describes the colors of the different planets' rays as they can be seen from the earth, and distinguishes the more or less harmonious angles at which the rays of the heavenly bodies may converge on the earth at any given moment. Despite all these factual truths, however, he questions the practical value of almost any binding conclusion that, as he phrases it, "astrologers are likely to read into" such facts.

[3]The Latin original, *De fundamentis astrologiae certioribus*, appeared in Prague in 1601. I quote from the commented translation by J. Bruce Brackenridge and Mary Ann Rossi, published in *The Proceedings of the American Philosophical Society* 123/2 (April 1979), pp. 85-116 [90].

Kepler's reservations are unequivocal. He rejects all claims he considers magical and all mystical cosmology (thesis 49: "nothing is more intolerable in this almost unique concern of certain astrologers than that they distribute with a certain infantile credulity twelve houses among seven planets"). He objects to any explanation that mystifies events for which there are simple reasons (thesis 65: "it is of the utmost importance to investigate the nature of winds, and very foolish to seek other causes of these [agricultural] events in the cardinal configurations"), and to all statements that seem to attribute volition or purpose to the heavenly bodies (thesis 75: "as if heaven might be said to be founded for afflicting mankind.").

What Kepler does acknowledge, and emphasize, is the unique impact of *harmony*, which he believes not to cause anything of its own accord, but to strengthen existing tendencies. Consonances occur in specific combinations and under certain geometric conditions of incoming rays. As Kepler explains the latter,

> there is some action [...] of heaven on the earth when it meets an intermediating ray of a star and then they come together on earth to form angles; thus the eight harmonic ratios will be expressed in the dimension of these angles, commonly called Aspects (thesis 38). [T]his influence does not come about because rays join to form any angle [; it] is manifested finally when the angle possesses a harmonic ratio (thesis 39). Now this faculty, which adds force to aspects, is not located in the stars themselves. For these aspects we are discussing touch upon the earth and are a pure condition flowing not from the formal motion of the stars, but from the accidental arrangement of pairs of stars with the earth. In the same way, then, that the spirit moving the body is located not in the object, but in the place where the idea of the object is displayed, just so is it necessary for this force, which makes the aspects powerful, to be located within that great globe earth, as it is in all the sublunary bodies. To be sure, every animate faculty is the image of God the Geometer in creation, and He is inspired to His task by this celestial geometry of aspects of harmony (thesis 40).

The angles whose force Kepler recognizes because of the strength of their *harmony* correspond to two geometric properties, the angular difference and the segment of the circle. Since the latter is expressed as a ratio, it can be translated into a musical interval. Here are the "Aspects," organized in accordance with the overtone series:[4]

[4]My diagram is inspired by that used in J. Bruce Brackenridge's comment on Kepler; see "On the More Certain Fundamentals of Astrology," p. 114.

FIGURE 30: Planetary angles, circle segments, and musical intervals

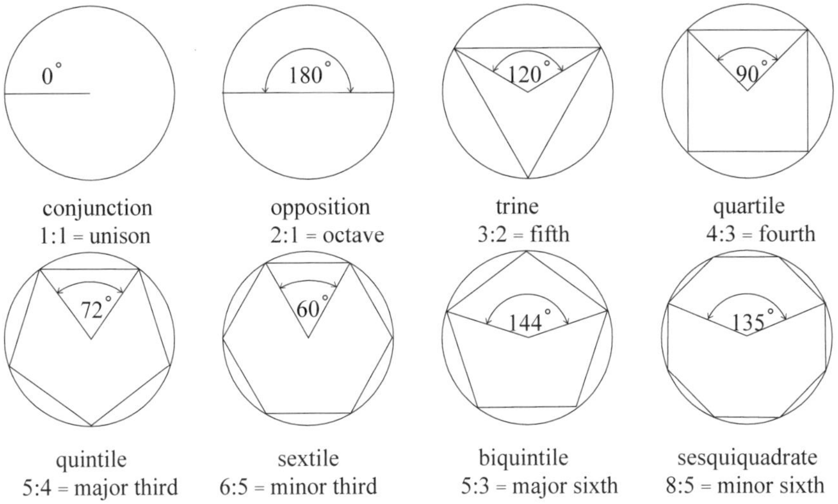

The strengthening influence of these consonant angles are felt by all living beings, Kepler claims. What inflection of well-being is helped along depends on the natural dispositions and momentary inclinations—of crops or persons. "Indeed the very same agitation, just as it gives those badly disposed over to destruction, just so it moves the stronger ones, either by reason of age or health, to great accomplishments. [...N]othing is clearly inevitable" (thesis 75).

Kepler openly scorned popular astrologers, whose pronouncements he considered at best mystifying and irrelevant, at worst irresponsible and harmful. In view of the few data he considered relevant and the many he deemed superstitious, it seems intriguing to investigate his attitude to personal birth horoscopes. As already mentioned in passing, he was convinced that the "twelve houses," determined by the zodiac sign rising over the eastern horizon at the moment of a person's birth, are overvalued and serve above all for charlatans' business. Hence he met the common belief that they represent different aspects of a person's life, and that the planets appearing in them must be indicative of personality or fate, with much skepticism.

Nonetheless, beginning as a young scholar and continuing into his late years, Kepler calculated innumerable personal horoscopes. It seems that at times he would have preferred not to be asked but was unable to forego this considerable source of income. His two famous horoscopes for Albrecht Count Wallenstein (1583-1634), the leading imperial general in the Thirty Years War who played a crucial role in Kepler's personal and professional life, are a case in point. In 1608, working on an anonymous person's request without knowledge of the general's identity, he makes cautious and often rather general statements. In 1625, now familiar with Wallenstein both as a public persona and as a personal acquaintance, he corrects and qualifies his earlier observations in interesting ways. Having specified the auspicious and dangerous periods over the initial five decades of Wallenstein's life, Kepler predicts an extremely inauspicious constellation involving Mars for the beginning of March 1634. While his early horoscope had continued way beyond this date, the second horoscope breaks off with the mention of the dangers looming at that moment over Wallenstein's life. Since the general would not yet be 50 at the time, this cessation might have warranted an explanation, but Kepler provides none. And he did not live to learn that on 25 February 1634, Wallenstein, who had finally and irreparably fallen out of favor with the emperor and sought individual agreements with the enemy, was surreptitiously murdered.

Kepler's two horoscopes for the notorious politician have been translated into various languages and attracted a wealth of commentary. Yet there seems to be no overriding consensus on what degree of credence their author himself accorded his own prognoses. Some commentators interpret the concurrence of Kepler's discontinued observations with Wallenstein's violent death as proof that astrologically-based predictions, as long as they are undertaken by persons of science with a knowledge as sophisticated as was Kepler's, are potent means of inference. Others argue that although Kepler had repeatedly warned the superstitious general not to take horoscopes as descriptions of what would inevitably occur, Wallenstein placed so much trust in them as to unknowingly conspire with fate and thus "attract" his death.

The dual question of what the stars or other harmonious constellations in the environment can foretell about earthly fates and in what way human lives can potentially partake (or exhibit aspects) of universal harmony, plays in various direct and indirect ways into Hesse's novel as well as Hindemith's opera.

Astrology in Hindemith's Opera

The many references to astrology in *Die Harmonie der Welt* are limited to the verbal level. While Hindemith does not seem to have sought ways to invent musical signifiers for this aspect, his libretto shows just how carefully he reflected upon Kepler's ambiguous stance within this field.

Act I is replete with allusions to the constellations, even beyond mother Kepler's moonlight activities and Emperor Rudolf's telescope experience. The opera opens with Tansur's brazen attempt to make money by linking political events to a recent comet ("Misfortune, suffering, the dire need with which the long-haired star threatens us ..."). Ulrich's first words correct the simplistic notions with which Tansur hopes to convince his gullible listeners, and introduce him as the assistant of one who is much more knowledgeable about the stars, but also more rational when claiming their impact on human affairs ("Comets wander in the distant heavens, generated spontaneously like the sun.") Wallenstein enters the plot with the request that Ulrich ask his master to cast his birth horoscope. Kepler's daughter questions the value of astrological prognosis when she asks her father concerning her mother's death: "Couldn't you read it in the stars?" Upon which Kepler admits: "All that is to be seen there is unclear, even to an untroubled creature," indicating that the preconceptions of the one doing the reading are so much part of the prognosis that the meaning of the observable constellations is never conclusive.

Steeped in sad memories, Kepler then intones his lament. Its reference to the weak human eyes that see nothing but what they expect casts additional doubt on the value of any human interpretation of celestial constellations. Ulrich, whom Kepler reminds of his duty to calculate Jupiter's orbit on the basis of Tycho Brahe's data, attempts to divert his master's attention from his laziness by producing an astrological prognosis of his own fate, a skill he claims to have learned by observing Kepler. Predictably, the young man finds only "greatness and fame" predestined in the stars, whereas Kepler, observing wryly that the figure is distorted "only by a couple of degrees," is surprised to see his assistant face "a strange fate as a soldier." When Ulrich indignantly rejects this thought, Kepler is quick to accede with a didactically aimed grin: "There you have the value of star-gazing."

(In his opera, Hindemith has Ulrich in fact end as a soldier. Considering his protagonist's emphatic distrust of horoscopes, this may seem surprising. One is reminded of the contradiction between the historical Kepler's accurate prognoses—e.g., of Wallenstein's death year—and his general discomfort with this side of his profession. Similarly, the operatic Kepler's request that Ulrich concentrate on Jupiter's course can be heard as a veiled prediction that the erstwhile astronomer's assistant will eventually be following the course of Wallenstein, whom Hindemith allegorized as Jupiter.)

In act II, Wallenstein speaks with confidence of the things he has seen "in the stars": his ascent to power, his unloved but rich wife's proximate death, and the imminence of independent wealth. In act IV, he asks Kepler for a new birth horoscope. Kepler reacts with embarrassment, cautioning that one should not set too much store by these calculations. It remains ambiguous whether he feels taken aback that at this height in his career, he is still not respected as a serious scientist, or whether his discomfort stems from the fact that he would rather not have to confirm what he had seen earlier but left unsaid: Wallenstein's premature and violent death. Yet Wallenstein will not be dissuaded from his conviction that the stars warn or encourage in time for those who care to listen to reap maximum benefit.

Three *dramatis personae* move in murky realms close to but not identical with astrology. When Ulrich, by the end of the opera a cynical soldier, mocks Kepler's ongoing search for universal harmony, sneering at anybody who believes in "heavenly book-keeping" in a world governed entirely by chance, he seems to confuse if not conflate astronomy and astrology. When mother Kepler fears that her son's attempts to probe ever deeper into the laws of astronomy may violate secrets that should remain concealed, she seems to assume that the cosmos is a being with thoughts and emotions, capable of feeling insulted or shamed. When Emperor Rudolf expresses his longing to slip from the phantom-world in which he lives as a sovereign "to the indefinable sphere of starbright, slumbering wisdom," and when the sight of the heavens with their seemingly orderly constellations prompts him to exclaim that "those laws show the worlds their paths," he treats the cosmos as a benign realm welcoming to humans. When the sudden abundance of luminous bodies made visible with the help of Kepler's telescope strikes him as "vile, mean, bad, and undignified," casting him into despair and momentary madness, he clearly attributes to the cosmos both volition ("Curse the demon that treacherously permits this") and a direct impact on human beings ("Instead of help, the final thrust into the dreadful abyss").

None of these three persons understands that Kepler, searching for manifestations of universal harmony, is seeking for the prototypical design in ideal (i.e., simple) proportions—a design that underlies all aspects of the divinely created universe, without any aspect therefore *causing* another. Hindemith's libretto attacks this mistaken view twice. In the moon rondo of act III, he has the lunar voices firmly reject all suggestions of an influence on earthly occurrences. Their statements climax in a declaration that not only makes human beings fully responsible for their actions, but also exposes the poor behavior their claims of "astral impact" try to excuse:

> I beg you, understand the moon correctly: It is not she who has planted evil in you, nor incited you to it. But if your doings seem to others weak and insincere, incompatible, incomprehensible, full of selfishness, indecisive, then her uncovering ice-light will penetrate your being without pity.

Similarly, the chorus of the Milky Way in act V declines any commitment to the inhabitants of planet earth:

> Neither our happiness nor the happiness of others avails such a sublime postulate; it does not lend itself to the service of any purposes, does not cause harm, does not provide benefit, does not prevent war and plague, danger and pain.

Soul and Society in *Die Harmonie der Welt*

In act II, Ulrich gains prominence as a force that counteracts Kepler's turn to the inner orbits of love and faith. As Kepler's assistant, he is first seen herding his master's students to the Linz Sunday service during which pastor Hizler will announce Kepler's exclusion from communion and church attendance. The public humiliation leads to this act's first climax, centered in the call for freedom of conscience—for which Ulrich alone among the churchgoers does not feel Kepler deserves support. It is also Ulrich who, through his clumsy effort at courting a childhood friend turned attractive young woman, introduces Susanna. She represents the second focus within act II: Kepler's beloved future second wife, with whom he is eager to share his professional and spiritual aims. In this sphere, Ulrich is openly antagonistic. He intercepts Susanna's guardian and, in the mistaken hope to ingratiate himself with the baron as a preferable bridegroom, maligns his master. He claims that Kepler is professionally unsuccessful and privately incompetent. He even alleges that Kepler may have had a hand in Barbara Kepler's death, as if trying to match the pastor's reproach of heresy with a hint at criminal behavior.

The scene in which the basis for both developments is laid is a scherzo in $\frac{7}{8}$ time, characterized by an almost uninterrupted pulse in drum, triangle, and cymbal with persistent irregular accents: $\underline{1}$ 2 3 $\underline{4}$ 5 6 7 | $\underline{1}$ 2 3 $\underline{4}$ 5 6 7. While its tangible raucousness may be superficially justified by the unruliness of students little inclined to Sunday duties, the "unharmonious" numerical ratio of 7 : 8 clearly serves Hindemith as a symbol for the intentional adversity the protagonist suffers in act I, which is so much more jarring than the trouble caused Kepler in act I by confused but not purposefully inimical persons.

We first learn about pastor Hizler's pronouncement regarding Kepler's exclusion from communion through the agitated comments and reflections of the congregation leaving Linz's Lutheran church. Kepler's students, the city and district representatives, and the women of the landed gentry can have known the new professor of mathematics only briefly at this time; yet they express unambiguous dismay at his outrageous treatment by their pastor. As if the need for defense against a representative of the Church prompted them to define their own opinions, they enumerate the qualities they believe to see in Kepler.

The society of Kepler's new home town initially couches its objections in a fugato.

EXAMPLE 19: The fugato subject of the dismayed congregation [11]

(Does the office of pastor really include
[acting as judge, bailiff, and provost?])

The subject's twelve entries are launched from the following tones: F♯ E♭ D F♯, F♯ E♭ A G A G♯ E F♯, thus showing a clear mooring on F♯. This tone in fact serves as a frame both in the four initial statements, which make up the instrumental segment, and in the eight remaining ones, which involve the chorus. At the same time, the strictly polyphonic composition is underscored by continuous drum and timpani rolls, whereby the timpani sustains the pitch E. With this pedal-note in mind, a new look at the subject reveals that E is indeed a pitch of secondary importance also in the melodic line (see the first measure in Ex. 19 above). One wonders whether Hindemith intended to link this scene, with its relevance for Kepler's experience of existential misery, to the tones of planet earth, *mi* and *fa*. The consecutive fourths following the initial seconds may then be heard as a brief allusion to the longer chain of fourths following the opening *mi-fa-mi* in the overture motif. There is surely *mi*sery: Pastor Hizler's claim that Kepler be considered a heretic is, after all, an allegation almost as serious as a charge of witchcraft. Yet it is not *fa*mine, spiritual hunger, that informs this scene; Kepler knows himself a seriously searching Christian who must find his inner sustenance in himself alone. Hence E is not joined by F, the pitch that in act I serves as a root for one needy person after another, but by F♯, which is slightly more remote.

The occurrences surrounding the pastor's discrimination against Kepler culminate in a grand chorus and ensemble in two segments, both led by Susanna. In the first, marked "slow and passionate," the churchgoers intone a solemn, mostly homophonic chorale in which they argue that such a treatment of a respected citizen is the Church's loss. While five soloists—Kepler, Hizler, Ulrich, the baron, and Susanna—sing five independent texts, the congregation at large accuses their pastor of being stubborn and sinful. The texture, which is reminiscent of a motet, provides an apt symbol of voices with individual arguments all coming together on a single claim: that for the freedom of conscience.

Benevolent versus Coercive Applications of Kepler's Harmony

Kepler's and Susanna's encounter in the latter half of act II climaxes in the future spouses' declarations of their respective and shared creed. Kepler's belief, which emphasizes

> the most profound order and magic of the laws as manifested in music, geometry, in the structure of plant and animal, the course of the stars, in essence, motion, relation, in all that has been created; which also includes the place, thought, deed, and destiny of human beings

is Pythagorean in every sense including the spiritual. The Greeks in Crotona would have agreed that the pervasiveness of majestic order must be understood as a manifestation of a divine fiat, the creation of a transcendent Mind. Susanna's own spirituality prompts her to translate Kepler's thoughts about harmony into precepts of loving-kindness. Only at the end of this act does Hindemith reveal her puzzlement over this suitor's unusual wooing: she sings about having won

> a crown of planets, a veil of comets, a castle built of suns. A universe set about me, an Olympus, fabricated from numbers [...]; an Eden of numbers, calculations, proportions, where formulas live instead of snakes, apples are harmless spheres. No flaming sword will drive us out; I will be happy to stay with Adam in his wisdom and stillness.

Two motifs assume the symbolic representation of the lovers' search for harmony. One, instrumental, accompanies Kepler's explicit proclamation of his spiritually informed professional goals;

EXAMPLE 20: The motif underlying Kepler's faith in universal harmony [15]

the other, primarily vocal and initiated by Susanna, explores the social application of universal harmony, especially with regard to people with whom it may be difficult to harmonize but to whom, she feels, one should extend particular affection and care.

EXAMPLE 21: Susanna's application of the concept of harmony [15 C-I]

(Should not we ourselves also achieve)

In musical structure and texture, Susanna's ardent declaration is unique. Not fashioned along the lines of any familiar motto, it combines various polyphonic techniques, primarily in the winds, with homophonic accompaniment in the *pizzicato*-playing strings. The initial measure of her vocal line is imitated so as to sound six times altogether.[5] From the second measure onward, it attracts a counterpoint played by the oboe, with which it engages in a dialogue reminiscent of Bachian *obbligato* arias.

EXAMPLE 22: Susanna's charity and its instrumental partner

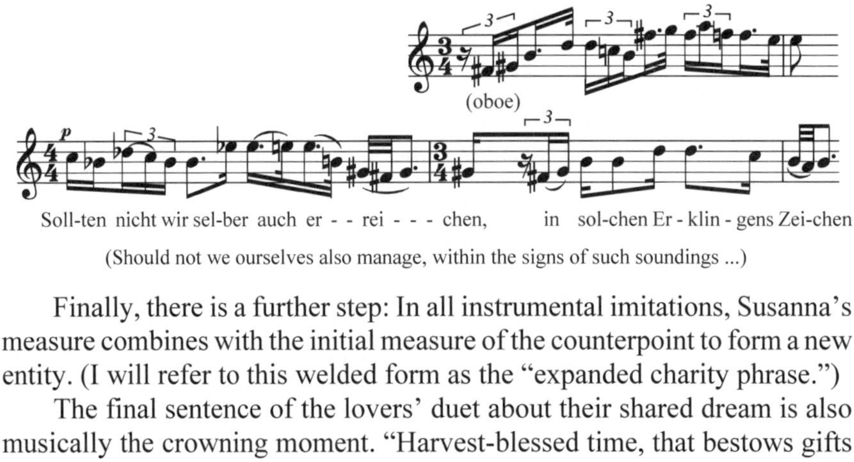

(Should not we ourselves also manage, within the signs of such soundings ...)

Finally, there is a further step: In all instrumental imitations, Susanna's measure combines with the initial measure of the counterpoint to form a new entity. (I will refer to this welded form as the "expanded charity phrase.")

The final sentence of the lovers' duet about their shared dream is also musically the crowning moment. "Harvest-blessed time, that bestows gifts upon us in infinite abundance and grants fulfilment of our deepest yearnings" sounds in four-part polyphony, whereby Susanna and Kepler set two new contrapuntal lines against the "expanded charity phrase" in the flute and the motif of Kepler's faith (cf. Ex. 20 above) in the trombones.

[5] 2 x Susanna, 2 x Kepler, 2 x instrumental; cf. Susanna at [15 C], imitated by Kepler/clar/bsn at [15 D], then consecutively by vc/db, Kepler/bsn/hr, Susanna/fl/Eh; by the flute at [15 F], and in the form of a 6-note incipit throughout the remainder of the future spouses' encounter.

Kepler's professional ethics (expressed in the question: "Should I not exhort the thinkers, artists, rulers of this earth to accommodate themselves within the confines of their work to that Word, so that through them, humankind will become conscious of the harmony of the world?") is balanced by Susanna's ethics of caring ("Should not we ourselves also achieve universal harmony, be close to our neighbors, understand them, better any evil, forgive, destroy hostility, propagate love?"). In conjunction, the complementary attitudes promise a powerful and equalized foundation for this marriage. I do not see Hindemith portraying Susanna as the lesser partner dominated by Kepler's single-minded zeal, as Winrich Hopp has recently argued.[6] It is Susanna, after all, who immediately grasps and invites her future husband to participate in a feasible human dimension of universal harmony. And it is hardly her fault that Kepler will later forget what he had once understood—to the point where, under the duress of his own disappointment while in Wallenstein's service, his worry about his family's welfare, and his concern about the spiritual degradation of his life's goal, he can no longer see what is there to console him.

In terms of the plotline, act IV can be regarded as a counterpoint of sorts to act II. Just as act II develops toward Kepler's marriage with the supportive Susanna, act IV heads toward his relationship with Wallenstein, whose attitude challenges Kepler to the core. Susanna stands in awe of the vision Kepler lays out before her inner eye; she becomes his spiritual equal when she generates her own vision of how the law of harmonious proportions must apply in the realm of human relationships. Wallenstein is as eminent in political and military matters as Kepler is in science and thus his equal in terms of achievements; yet while he confesses his admiration for Kepler's *Harmonices mundi*, he clearly intends to exploit its insights exclusively toward his own ends.

How much this strong character, who will by the end of this act have offered Kepler an income and a refuge from the rampant religious persecution, is a danger to the protagonist's integrity, is made apparent in a detail of the underlying music's thematic material. The conversation in [23] juxtaposes Wallenstein's faith in horoscopes with Kepler's reservations about them and prepares for the general's self-serving adaptation of the idea of universal harmony. The music turns here on an instrumental motif that is a conspicuous take on—and distortion of—the vocal motif with which Susanna, in act II, sang of her social and charitable understanding of the idea of a harmonious world.

[6]See Winrich Hopp, "Von der harmonikalen Fehlgründung lebenspraktischen Glücks: Zur Sinfonie und Oper *Die Harmonie der Welt*," in *Hindemith-Jahrbuch* 2002/xxxi: 19-40 [28].

EXAMPLE 23: Wallenstein's and Susanna's adaptation of universal harmony

While Susanna focuses on what Kepler's concept of universal harmony must mean for mutuality among humans and what qualities of empathy and tolerance should therefore be her own goals, Wallenstein interprets the same idea as a scientific vindication for his desire to impose order by force from above. Kepler attempts to caution against such a misuse of a philosophical paradigm, but his critical independence is by now bridled as a result of Wallenstein's offer to ease his financial situation. When Kepler justifies his willingness to become Wallenstein's mathematician with constrained assertions about the need to explore the possibility of uniting "idea and usage," he reaches the lowest point in the betrayal of his life's goal.

In act V, we find Kepler sick in a Regensburg inn. As he understands that death is approaching and that he will not see Susanna or his children ever again, he sadly admits to himself that he has spent himself seeking to establish a semblance of earthly harmony in the wrong places while neglecting its purest manifestation. To his words,

> I was driven to rouse it where it did not exist and, disbelieving, to flee from where in the closest intimacy it seemed to offer itself to me,

his vocal contour quotes not only the by now familiar first measure of Susanna's harmonic creed, but the entire "expanded charity phrase." It may be too late for a last embrace in the mortal realm, but as Hindemith's music assures us, Kepler in his last living moments achieves the clarity of vision that allows him to harmonize his own intellectual concept of the consonances in the universe with Susanna's compassionate adaptation.

Transtemporal Development of a Soul: Knecht's Five Lives

The biographies of the man whom readers first meet in his incarnation as an exceptional *magister ludi* receive different interpretations from the fictional chronicler and from Hesse. The 25th-century narrator explains all earlier life stories as playful exercises at conjecture, homework assigned to Castalian students during their years of free study. Surviving in them, he says, is "a remnant of the ancient Asian doctrine of reincarnation and the transmigration of souls." His assessment of this habit, which the Castalia of his time has apparently cast off, is a mixture of amusement and wonder:

> All teachers and students were familiar with the concept that their present existence might have been preceded by others, in other bodies, at other times, under other conditions. To be sure they did not believe this in any strict sense; there was no element of dogma in the idea. Rather, it was [...] a game for the imaginative faculties, to conceive of oneself in different conditions and surroundings. In writing such Lives students [...] learned to regard their own persons as masks, as the transitory garb of entelechy (p. 114).

When Hesse explained the Lives—which, as the sketch from 1931 shows, were the cell from which the whole *Glass Bead Game* grew—he left it open what concept of rebirth he had in mind.

> The imagery that ignited in me the first spark was that of reincarnation as an expression of the stable element within the flow, the continuity of tradition and of the life of the mind in general. One day, many a year before I began the attempt of writing this down, I had the vision of an individual but transtemporal biography: I imagined a man who in several rebirths experiences the great epochs of human history. What remains of this original intention is, as you see, the sequence of Knecht's lives.[7]

The threads that weave their way through the five lives make it seem likely that Hesse understood reincarnation in the sense of Arthur Schopenhauer's "constant rebirths" that

> constitute the succession of the life-dreams of a will which in itself is indestructible, until, instructed and improved by so much and such various successive knowledge in a constantly new form, it abolishes or abrogates itself. [...] not metemphychosis, but a peculiar palingenesis.[8]

[7]Transl. from a letter of January 1955 to Rudolf Pannwitz; reprinted in *Materialien* I, p. 294.

[8]A. Schopenhauer, "On Death and its Relation to the Indestructibility of our True Nature," in *The World as Will and Idea*, Suppl. to Book IV (London: Paul, Trech & Co, 1906), p. 300.

In the first Life, Knecht grows up in some indistinct prehistoric era. A boy in a matriarchal forest community, he becomes the apprentice, son-in-law, and successor of the local rainmaker.[9] As such he is a revered member of his clan, but also somewhat set apart by his extraordinary wisdom and skill. He lives in harmony with the forces of nature and learns how to fear them less by understanding them better, how to read them, and sometimes influence them. Yet there comes the year in which no rain falls and his people are threatened with starvation. All Knecht's rituals and incantations remain unanswered. Believing that the gods must be dissatisfied with him and can only be pacified if he offers them his life, he sacrifices himself.

Joseph Knecht's second incarnation in Hesse's sketch of 1931 became the third of the *magister ludi*'s posthumously published fictional autobiographies. Like "The Rainmaker," "Indian Life" is mythical rather than historically specific. Its locations—hills, forests, and rural villages near a capital —remain unnamed; its time is vaguely referred to as "the golden age of India."[10] The protagonist Dasa (Sanskrit for "servant" and thus a translation of "Knecht") is born a prince. Threatened by the ambitions of a stepmother and stepbrother, he is led into safe anonymity by a kindly Brahman, grows up among herdsmen, and only comes into contact with his father's court in

[9]In May 1934, Hesse wrote to H.C. Bodmer: "When Ninon [his wife] asked me what I was 20,000 years ago, I often told her that I then was a weather maker. In order to make it even clearer to her, and for other reasons as well, I wrote the thing down in the course of last year. Now it has been printed and I can herewith send it to you." Transl. from *Materialien* I, p. 87.

[10] The term "golden age of India" allows for two readings, one regarding the political and the other, the spiritual history of her peoples. In political contexts, the phrase points to the time ca. 320-500 C.E. when the emperors of the Gupta dynasty reunited northern India and expanded India's territory to include what is now Afghanistan as well as the Vindhya mountains in the south. Under the Guptas, the arts, literature, mathematics, philosophy, and science flourished. Many Hindu temples stem from this time, as does India's renowned system of Ayurvedic medicine.

More likely in Hesse's context is the religious connotation, which describes an era ca. 1400-600 B.C.E. It began when groups of a warrior people, who called themselves *arya* (nobles), who tended sheep, goats, cows, and horses, and who measured their wealth in herds of cattle, left their homes in central Asia and moved south. On the subcontinent, they came upon the Dravidians, who enjoyed an advanced civilization, grew crops, and had begun living in towns. The Aryans gradually conquered the Dravidians, drove many of them southward, and eventually extended their rule over most of present-day India. In the course of several centuries, they established a caste system in which the Brahmans (priests) composed the highest rank and the Shudras (possibly the descendants of the Dravidians) counted as the lowest. The Brahmans perfected Sanskrit, the language of the Aryans, conducted elaborate rituals and sacrifices, and passed sacred knowledge from one generation to another. The earliest known Hindu scriptures, the *Vedas*, were composed from about 1400 B.C.E. onward, and the sacred writings of Hinduism, the *Upanishads*, appeared between 800 and 600 B.C.E.

the capacity of a deliverer of butter for sacrificial fires. A yogi whom he sees meditating in a forest speaks to his spirit, but a beautiful young woman entraps his senses. Dasa marries, finds pleasure and deception, and only then returns to the yogi. Having asked to be taught what Maya is and what "sham" means, he is granted a vision of his potential future. The magic dream that opens his eyes lasts only minutes in real life: he experiences succeeding his slain stepbrother as king, being coerced into waging a war, and finally failing, finding those he most loved slaughtered by the enemy. Awaking, he recognizes that this ostensibly busy and dread-filled life is really nothing, that it has no reality. This is Maya. Struck by the insight, he renounces the world and follows the holy man as a disciple, on the path toward an experience of cosmic Oneness.

Knecht's third incarnation in chronological order is somewhat more specific. Named Josephus Famulus (the Latin translation of his name), he lives in Gaza at the time of the early Christian desert fathers, in the 4th century, or even more precisely, during the later years of St. Hilarion, who died in 372. An adult convert to Christianity, Josephus is exceedingly pious but also full of doubts about his worth. After many years of living in a desert hermitage and listening to people confess their real and imagined sins, he despairs of what he senses as his spiritual barrenness. One day, no longer able to endure the inner torment, he leaves his refuge to seek out another anchorite and father confessor, Dion, in the hope that the more experienced man may fortify his faith. Significantly though, Father Dion himself is filled with a similar insecurity. Both men's lives are weighed down by the notion that human sinfulness is intrinsic and insuperable. Yet the older hermit masters his own qualms, Josephus finds the desired mentor and is gradually healed. He serves Dion and becomes his successor.

The 18th-century Swabian life is different from the preceding ones for three reasons: it is rather precisely dated (Knecht is born in 1707 or 1709), it has remained unfinished and was published separately,[11] and it gets much more attention within the Castalian story than does any of the other three.[12]

[11]Though incomplete, this life is by far the most extensive. In the first of two drafts, which compares best with the other lives as it matches their third-person mode, it covers 70 pages: 69 pages of narrative followed by 1 page of notes for the intended continuation. This dwarfs the length of each of the completed biographies—42 pages ("The Rainmaker"), 38 pages ("The Father Confessor"), and 38 pages ("The Indian Life") respectively.

[12]Of the autobiographies the chronicler includes in his alleged publication, only "The Indian Life" is mentioned in the Castalian story, with a single clause on p. 115. The Swabian Life, by contrast, is described in detail on pp. 115-116 ("We know from anecdotes and letters ..."), becomes a topic in one of Knecht's conversations with Father Jacobus (see pp. 164-167), and recurs again in Knecht's thoughts about his improved view of Castalian history (p. 174).

This time, Knecht grows up between a music-and-nature-loving father and a pietist, rigid, and socially ambitious mother. He passes the exam that grants him free schooling in Württemberg's demanding education system for future Lutheran clergy, attends the cloister schools at Denkendorf and Maulbronn (Kepler's Adelberg and Maulbronn, the Castalian Knecht's Eschholz) and finally studies at the Tübingen seminary (where Kepler also completed his education, and which is fictionalized with many of its geographic features in the Castalian Waldzell). All through his education, he is torn between his pledge to become a theologian and his inner vocation to music. Though full of doubts, he becomes a pastor, but finds no fulfilment. When he is already around 50 years old, he discovers the music of the recently deceased J.S. Bach, which much more than all theological carping seems to him capable of fulfilling his longing for harmony and a "service of the highest." He resigns the pulpit and spends the remainder of his life as an obscure organist.

All four "pre-births" of the Castalian *magister ludi* chime with the idea of a soul's development along recurring issues. These issues reveal themselves on close inspection as related to those that shape the protagonist's 23rd-century life story. Moreover, they are also instantiations of the themes and motifs ascertained in the discursive glass bead game threaded through the compound work, in Joseph Knecht's first *ludus solemnis* as I conjectured it on the basis of the "Chinese house-building" image and in Hesse's sequential allusions to the *I Ching*.

In what thus amounts to a fifth dimension of Hesse's presentation of these ideas, one can distinguish once again five conceptual complexes: two themes and three motifs.

1 The first thematic complex could be summarized as "rift / synthesis." At the outset, it is articulated as the contrast between the intellectual depravity of the Age of the Feuilleton and the sober and sophisticated idealism of the World of the Mind. Repeatedly taken up in the further movements of the discursive glass bead game, it crystallizes as the dichotomy between the reality of distance and the ideal of harmonization. In the *I Ching* thread, it informs the insight that aloofness is a blemish, since the ultimate goal is the higher unity between mind and nature, to be achieved both in contemplation and in action.

- In the prehistoric life, we find a simple precursor of the distance between the world of mind and that of nature when Knecht the rainmaker discovers that in his community, "men of thought gave offense and aroused the repugnance of their fellows. They might be valued at a distance and called on in emergencies, but others neither love them nor accept them, rather give them a wide berth" (p. 465).

The distrust natural men feel toward wise men expresses itself in the fact "that the sick and unfortunate are far more receptive to traditional magic spells and exorcisms than to sensible advice; that people more readily accept affliction and outward penances than the task of changing themselves" (pp. 465-66). In his bitter moments, Knecht in turn has a poor opinion of the majority among his contemporaries who have not developed either their rational faculties and judgment or their self-possession. He calls them "weak, selfish, and cowardly creatures ..." (p. 466).

- In the incarnation as the Indian prince-herdsman Dasa, the topic is developed in several steps. A first dichotomy juxtaposes herdsmen with settlers (thus repeating the mythical victory of Aryans over Dravidians). The former are idealized as the separated few, who live removed from the world of material goods and status concerns, in contrast to the self-important many, the settled people with their noise and pomp (p. 526). Correspondingly, Dasa the herdsman is susceptible to a yogi, whereas as a settled householder he succumbs to carnal craving, discontent, and eventually even to jealousy-driven murder. A similar contrast, occurring in what turns out to be nothing but a bad dream, exists between Dasa the reinstated king and the former herdsman's wife become queen: her chief pleasures lie in festivals, parades, and pomp, while he prefers the joys of his garden and is attracted by learning (p. 542). The Castalian version of the dichotomy shines through in particular when we learn that as a king, Dasa compiles a library of sacred texts, invites the foremost scholars and thinkers to conduct disputations on sacred subjects, and regards everything that pertains to sensual life and "human nature" as both strange and dubious, at once touching and ridiculous. The forest yogi, presenting a third stage in the world–nature contrast, has taken the exclusive concentration on mind and spiritual purgation to the extreme: he "had plunged through the surface of the world, through the superficial world, into the ground of being, into the secret of all things. He had broken through and thrown off the magical net of the senses, the play of light, sound, color, and sensation, and lived secure in the essential and unchanging" (p. 524).
- In the Christian anchorites' environment, it is no longer an individual but a select group that is contrasted with the unwashed masses. Hermits "seemed to have assumed the duty of making up for the pleasure-seeking, brutality, and sensuality of many past and future ages by engendering a mighty surge of enthusiasm and devotion, an ecstatic excess of renunciation" (p. 489).

- The Swabian Knecht is the first to show signs of a reverse opinion. During his school years, he perceives the "world of learning" and the "republic of scholarship" as sadly juxtaposed to "the freedom, innocence, and natural aimlessness of childhood" (*Fourth Life*, p. 148). Education for the pulpit happens in various "temples of knowledge" (p. 150), but it is music that provides "food for the soul." Even within the religious field, Knecht perceives the erudite among his fellow students and teachers as deficient in comparison to those who are capable of a direct, childlike communion with God.

2 The second thematic complex combines the ideas summarized in the Castalian chronicler's introduction under "ancient Chinese wisdom," the *I Ching* motifs "power of smallness" and "learning with respect," the pervasive ideal of "service," and the dialectic of hierarchy and individuality, which appears both in the fourth stage of the discursive glass bead game and in the "Chinese House" game.

- Knecht the rainmaker spends his life in service, first to his master, then to his profession and, through it, to his community and all forms of life in his area. He understands that "a man is the more useful, the more his life and thinking is turned toward matters of the spirit, matters that go beyond the personal realm, the more he has learned to venerate, observe, worship, serve, and sacrifice" (p. 465). In a blend of humility with self-respect he realizes that he is "an unknown link in a chain, but a link as indispensable as any other" (p. 469).
- The Indian Dasa also submits to a mentor before turning to the ideal itself. When he first lays eyes on the yogi, he feels that he must serve him. When he returns to the forest after many years among the settled people, he immediately sets out to repair the ascetic's hut, collect fruit for him, and fill his water gourd. "Dasa lived with him like a servant in the presence of a nobleman"(p. 533). When finding the path to enlightenment turns out to be all but easy, he is able to articulate that at least, "obeying and serving were better and far easier, seemlier, and far more harmless, than commanding and taking responsibility"(p. 557).
- The Christian desert hermit Josephus Famulus distinguishes himself through his extraordinary gift of serving others: his gift of listening, patience, and silence, his non-judgmental receptivity and the ease with which he treats all penitents with deference. When his despair over his perceived emotional dryness tempts him to commit suicide, he reflects whether even this deed might not be committed in the spirit of a sacrifice, as in situations where "the chosen man of the tribe gave up his life for the general welfare, often by his own hand"

(p. 497). Later on, in his exchange with Father Dion, Josephus experiences a third form: a "service through smallness." Both he and the older hermit confessor suffer from being thought saintly while they in fact feel tired of their work and have begun to doubt its value. In this situation, the one's distress causes the other's commiseration and thus becomes a healing and fountain of strength for both.

- The Swabian Knecht is distressed that among the clergymen of his day—servants of God by definition—many seem almost exclusively concerned with their own interpretations of the "right" understanding of God's Word. Knecht is humble by nature and has no doubt about wanting to serve, but does not dare embark on the path that his instincts tell him is the one to which he is called. Misled by his mother's prejudice that music, like people, exists primarily to sing of God's glory, he forces himself to ignore his inner calling and follow an unloved career until it is nearly too late to realize his soul's true potential.
- In each of the Castalian Knecht's pre-existences, all the learning that ultimately matters is absorbed not from books and formulas but from mentors, less through questions than through silent observation, in a spirit of reverence rather than of eager impatience. Each of the lives takes Knecht toward a position in which he is considered outstanding and exemplary; yet he desires modesty and honesty, so that he may feel true to himself. In the prehistoric era, he relinquishes his position by giving his life; in the Indian story, narrator and reader are excluded from the possible qualms of the would-be yogi. In the Gaza life, his move to abandon the exposed position of a saintly hermit is explicit though deflected; in the Swabian life, it is consummated—with the implication that the "obscure organist" has found a contentment and truth that was not available to the respected pastor.

3 The first of three recurring motifs in the prior lives is meditation, in connection with its main goal: serenity and the enlightened recognition that the rift between mind and nature concerns only life's surface. Knecht the rainmaker commands a degree of composure that allows him to face life's disturbing secrets better than his fellow villagers. This poise, based on contemplation in its rudimentary form, is an early stage of the more deliberate meditative exercises described in later lives. The Indian Dasa discovers two kinds of serenity, the instinctual and the devotional one, this and that side of humankind's social life. As a herdsman, he perceives nature as serene, undisturbed by people's greed and passions. The yogi whose path he eventually decides to share is described as surrounded by an

"aura of holiness", "a magic circle of dignity, a flame of concentrated intensity and a wave of radiant yoga energies," as emanating "waves and rays in the midst of which he sat enthroned like a moon" (p. 523). Whereas in the serenity of a life in nature, sensual and social ambitions are largely dormant for lack of stimulus, in the world of the yogi they are purposefully and painstakingly excluded, to the point where the yogi lives in a dimension all of his own. "The whole world might be no more than a breath of wind playing over the surface, a ripple of waves over unknown depths." In the father confessor's life, meditation is captured in the act of prayer, yet proves of limited power over the hermit's growing despair and sense of failure. Serenity appears here like a natural state that is gradually eroded by a life not nourishing its basis. In the Swabian life, a clandestinely observed act of meditation—the admired local rector's prayer in a lonely attic, which visibly relieves the devout man from his sorrows and gives him tranquility—serves as a calling for young Knecht. Yet once again, redemptive prayer is a gift. Knecht's education as a clergyman does not seem to offer him help as to how such absorption and healing may be learned; his theological instruction leads rather in the opposite direction.

4 The second motif is that of ritual and art. The Castalian chronicler describes the ceremonial performances during the heyday of the glass bead game as ideal marriages of ritual and art. As I have shown in a previous chapter, the Confucian ideas behind the tradition of Chinese house-building favored edifices informed by and centered around ancestral worship and rituals of *feng shui* and astrology, but also invited music, poetry, and various aspects of the visual arts to contribute—albeit in their highly ritualized rather than their sensual forms. The same holds true for art as practiced in Castalia: it serves as a treasure trove of culture, to be studied for edification and spiritual delight but not to be practiced for emotional wallowing or a striving toward fame.

The first and the last of the four "pre-incarnations" are most explicit in regard to this motif. Knecht the rainmaker is highly skilled in using rituals and simple forms of art for his office. His regular tasks rely on "a great heritage of tradition and experience, the sum total of man's knowledge of nature and even more [...] a dense system of experiences, observations, instincts, and habits of investigation" (p. 455). When a shower of shooting stars seems to obliterate the familiar firmament and his community is mad with fear, he is able to dispel terror and restore order in most hearts with the help of music, dance, and mime. Leading the villagers in rhythmic shouting, clapping, stomping and incantations, he achieves amazing results. As Hesse's commenting voice observes: "Each now fitted into his place in the orderly chorus of the multitude, keeping to the rhythm of the exorcistic ceremony. Many mysterious powers are present in such a rite. Its greatest comfort is its uniformity, confirming the sense of community; its infallible medicine meter and order, rhythm and music. [...] In their fear and weakness men met the disorder of the sky with order and religious concord. Even before the rain of stars began to slacken, the miracle had taken place, [...] and by the time the sky seemed slowly to be quieting down and recovering, all the dead-tired penitents had the redeeming feeling that their worship had placated the powers and restored order in the heavens" (p. 480). For the 18th-century Swabian Knecht, things may seem reversed: music is his true love, perhaps even his secret religion; it is the wished-for end and not, as his pious mother preferred, only a means. To the boy, the most attractive part of hymn singing is "the harmony, the interweaving of parts"; the most awe-inspiring component of Sunday service is the sound of the organ. In later life, he discovers that Bach's music expresses true religious devotion better than any words do. "Music has a profound, magical healing power, and even more than the other arts it can serve as a substitute for nature. It was music that saved Knecht's soul from intellectual aridity." (*Fourth Life*, p.149)

5 The third motif is the "calling" or "awakening," juxtaposed but usually related to doubt, as well as the concept of the successor who will ultimately fulfill the calling. Knecht the rainmaker experiences his main calling during the night watch when his master instruct him about life's superior orderliness: after the older rainmaker's death, his spirit will fly to the moon, but return to dwell in Knecht's next-born son, who will in turn one day become Knecht's successor. At the sight of the meditating yogi, young Dasa feels called to a life he would never have known to seek; when he stumbles across the same scene again many years and bitter experiences later, he knows that he has been "guided by the voices in his soul" and feels "as if he had just awakened" (p. 532). After a long trial, the yogi silently accepts him as his apprentice and eventual successor. "The chances were that this young man had already been wakened once or several times before, and had breathed a mouthful of reality, for otherwise he would not have come here and stayed so long. But now he seemed to have been properly awakened and become ripe for setting out on the long journey" (p.557). In Gaza, the two desert fathers feel secretly called

to one another and become instrumental in each other's awakening. Josephus lives on as Dion's appointed "son" and successor (p. 516). When the Swabian Knecht witnesses with awe the local rector's redemptive prayer and feels his eyes opened to true piety, he interprets this experience as a calling. In his studies, a few impressive teachers cause small degrees of further awakening, but Knecht's path in their footsteps is hampered by the pervasive strife in official theology. His final life phase as an organist, while said to have given him quiet happiness, remains pale (and perhaps for that reason incomplete) as it is neither fostered by a living mentor nor crowned by a successor.

Doubt plays a role in all these lives. Whenever his intimate connection with the world of plants and stars seems broken, the prehistoric Knecht is troubled by the thought that he may not be a real rainmaker; and yet, he never thinks of resigning or withdrawing. The 4th-century desert father accuses himself of spiritual pride, coldness and lovelessness, and leaves his function and his ministry in order to seek rather than to administer spiritual fortification. The 18th-century Knecht is one in a group of young men destined to the clergy who feel sorely dismayed: "every experience and insight that reinforced them in their piety shook their confidence in their vocation" and "the teachings of the church were at odds with secular philosophy and with the religious needs of the people" (*Fourth Life*, p. 171). He ultimately leaves his profession for another that better answers his spiritual longings.

The five thematic components of Hesse's *ludus* thus appear on five levels: philosophically in the interpretations of the *I Ching*, culturally embedded in the Confucian principles of traditional Chinese house-building, discursively verbalized in the movements of the glass-bead-game thread, psychogenetically in the lives prefiguring (or retro-projected from) the Castalian Knecht's experience, and finally in the extraordinary *magister ludi*'s destiny. In him, all themes and motifs climax. He feels most acutely the rift between the two worlds, longs for harmony, and is willing to sacrifice his career and even his life for it. He not only preaches but lives the idea of service and essential humility without therefore losing his emotional anchoring; he also learns from examples rather than from doctrines. Meditation sustains him, music is his greatest love, the glass bead game captivates him because it combines art with ritual, but doubt in the chosen path and a series of awakening experiences prompt his decision to abandon his honorable position for a humble but honest one, where he hopes to train a successor who will complete the task of harmonization.

Part VII
ART AS A HYMN TO UNIVERSAL HARMONY

Plate 7a: Construction of a human head.
Sketch with triangulation according to the painting school of Beuron

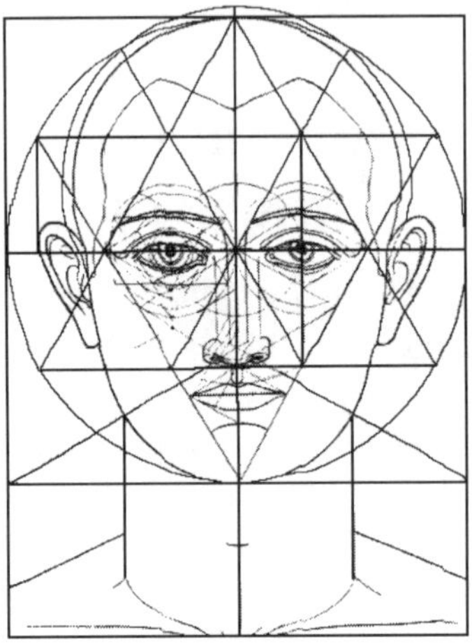

Plate 7b: Sectional view of the Church of St. Petronius in Bologna.
Etching with triangulation by Floriano Abrosini (1592)

Human Art: Imitating Nature's Proportions

The Roman architect Vitruvius Pollio (84-14 B.C.E.) compiled Greek writings about architecture, both to make their insights available in the Roman Empire and to convince his contemporaries how central proportions ("good measures," as Albrecht Dürer would later translate the word) had been to the Greeks. Although those for whom it was intended ignored it, Vitruvius's ten-volume compendium, *De architectura libri decem*, proved instrumental in preserving Greek knowledge for posterity. The Romans, grappling at this time with questions of brick-wall stability and vaults, were uninterested in the niceties of column spacing and other questions of internal structuring. But Vitruvius's writings survived his epoch and came to fruition in the Middle Ages.

Essential in the context of a study about "the musical order of the world" is the fact that Vitruvius links architectural requirements directly to the ratios defining musical intervals. In the fourth chapter of Book V he includes excerpts of Pythagorean music theory. They are not flawless in their understanding, since this was decidedly not the author's field of expertise, but they prove his essential grasp of the common ground. Vitruvius gives a thorough description of the tonal system, demands that all architects have a good training in the fundamentals of music, and tries to equip the designers and builders of his time with all that should enable them to create edifices that are in consonance with the universal laws of harmony. His reasons are both pragmatic and spiritual: he applies harmonic laws to solve problems of acoustics in theaters as well as to build temples that will provide a perfectly built space in which to worship the gods. The former is aptly documented in Book V of *De architectura*, where he states:

> Hence the ancient architects, following in the footsteps of nature, perfected the ascending rows of seats in theatres from their investigations of the ascending voice, and, by means of the canonical theory of the mathematicians and that of the musicians, endeavoured to make every voice uttered on the stage come with greater clearness and sweetness to the ears of the audience. For just as musical instruments are brought to perfection of clearness in the sound of their strings by means of bronze plates or horn, so the ancients devised methods of

increasing the power of the voice in theatres through the application of harmonics.[1]

In an attempt to validate the last claim, Hans Kayser examined the three great temples in the southern Italian (erstwhile Greek-inhabited) town of Paestum.[2] Measuring the three-dimensional ratios within single halls, the proportions of adjacent rooms, the angles and areas of friezes, and the radius, height, and spacing of columns, he concludes that all measurable entities relate to one another in single-digit ratios. In other words: every aspect in the temples' design exhibits proportional harmony with every other.

Fifteen centuries later, another Roman architect moved in Vitruvius's footsteps. In 1452, Leon Battista Alberti (1407-1472) completed a treatise of his own, reverentially entitled *Ten Books of Architecture* with a bow to Vitruvius. In the fifth chapter of Book IX, he explains with regard to the three-dimensional proportions of rooms and buildings that

> The Rule of these Proportions is best gathered from those Things in which we find Nature herself to be most complete and admirable; and indeed I am every day more and more convinced of the Truth of Pythagoras's Saying, that Nature is sure to act consistently, and with a constant Analogy in all her Operations: From whence I conclude that the same Numbers, by means of which the Agreement of Sounds affects our Ears with Delight, are the very same which please our Eyes and Mind. We shall therefore borrow all our Rules for the Finishing [of] our Proportions from the Musicians, who are the greatest Masters of this Sort of Numbers, and of those particular Things wherein Nature shows herself most excellent and complete.[3]

During the Middle Ages, the guild of builders perpetuated the knowledge of harmonious proportions in sacred spaces. Known in German to this day under the deceptively modest-sounding term "Bauhütte" [builders' workshop, literally: builders' hut], with "Dombauhütte" designating the august brotherhood of artists and artisans entrusted with the building of cathedrals, these guilds developed manuals, which they handed down through generations. The only surviving example is the "Bauhüttenbuch" (builder's portfolio) of Villard de Honnecourt, dated ca. 1230-35. It contains

[1]Vitruvius Pollio, *The Ten Books on Architecture*, ed. Morris Hicky Morgan (New York: Dover, 1960), Book V, Chapter 3, paragraph 8.

[2]Hans Kayser, *Paestum: Die Nomoi der drei altgriechischen Tempel* (Heidelberg: L. Schneider, 1958).

[3]Leon Battista Alberti, *Ten books on architecture*, ed. by Joseph Rykwert as a complete reprint from the 1755 edition (London: Tiranti, 1965), p. 196.

a schematic drawing apparently intended to solve with one comprehensive diagram several of the common tasks of spatial proportion. Hardly surprising at this point of the survey of harmonic thinking, the diagram can be shown to exist as part of the geometric realization of the *tetraktys*; it is also a mirrored double of the helicon. (The figure's height-to-width ratio is different in each case, but this proves irrelevant to the harmonic proportions for which it served.)

FIGURE 31: Drawing from an early 13th-century builder's portfolio, related to the *tetraktys* and the helicon

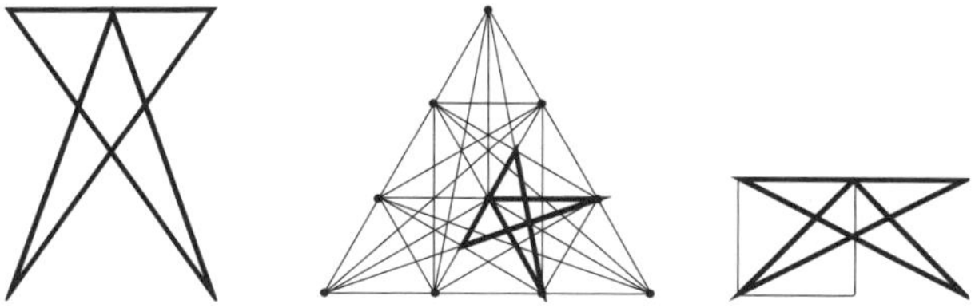

In a 1946 essay devoted entirely to Villard's diagram (above left), Hans Kayser called it "a canon for harmonious subdivision," showing how it can be used to yield many more than only the basic ratios. The process resembles that which the Pythagoreans are believed to have used when constructing the helicon:

- the diagram is fitted to a rectangle;
- two horizontal lines are drawn through the points where the diagonals intersect (A-A/B-B);
- these horizontals' end points are linked to the upper center (X);
- the linking lines generate new vertices;
- through these, new horizontals are drawn, whose end points are in turn linked to X.

This process, which could be repeated at infinitum, yields the entirety of simple ratios—musically speaking: all the consonant intervals of a scale.

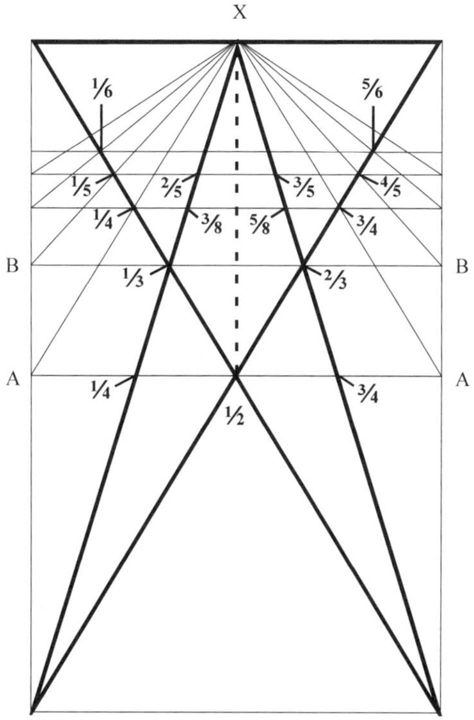

FIGURE 32: Villard de Honnecourt's diagram, a secret canon for harmonious subdivision

How universal both the thinking in harmonious proportions and the tools to visualize these were, can be seen in two examples originating outside the Western tradition. The famous Arab mathematician Ibn-al-Haitham (died 1038) used a diagram that is related both to the ancient Greek helicon and to the 13th-century builder's drawing. Indian artisans were even earlier in applying a variant of the same measuring tool.[4]

FIGURE 33: Arab and Indian variants of the diagram

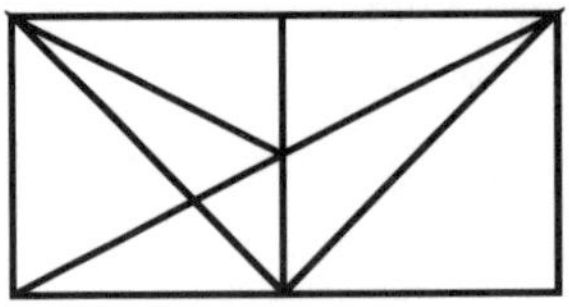

Since that early time, many other relationships between spatial proportions and musical harmony have been discovered. In 1888, the German acoustician A. Eichhorn wrote a treatise on "the acoustics of large halls according to Greek theory" in which he proved that ancient cultures knew to build their rooms in the proportions corresponding to a musical triad (i.e., in measurements relating to one another as 3 : 4 : 5, or the fifth, octave, and major tenth/third of an assumed fundamental). The organ builder Ernst Schieß collected evidence from acoustically successful halls all over Europe and testified that all of them conform to simple harmonic proportions.[5] Moreover, not only large rooms resonate better if their measurements correspond with those of musical consonances; the same is true for the bodies of instruments, as Hans Kayser has demonstrated for the specific shape and dimensions of the violin.[6]

When planning cathedrals, medieval architects favored geometric shapes built on the basis of the two leading numbers of Christian symbolism: THREE (for the Trinity) and FIVE (for the wounds of Christ, or the words in the expression "Jesus Christ God's Son Savior"). For reasons of perfection, only

[4]Ibn-al-Haitham's drawing is reproduced in A. P. Juschkewitsch, *Geschichte der Mathematik im Mittelalter* (Leipzig 1964), p. 281; the Indian variant is given, together with an application, in Rudolf Haase, *Harmonikale Synthese* (Vienna: Lafite, 1980), p. 86.

[5]Quoted in Hans Kayser, *Akróasis: Die Lehre von der Harmonik der Welt* (Basel: Benno Schwabe Verlag, 1946), p. 61.

[6]Hans Kayser, *Die Form der Geige, aus den Gesetzen der Töne gedeutet* (Zurich: Occident, 1947).

equilateral triangles were eligible, and these were used frequently. As Charles Hummel has shown in his study of the masters of Chartres Cathedral and their Pythagorean thinking, the vault is designed with the help of a pentagram inscribed into a circle (Fig. 34), while three equilateral triangles fitted to one another define the remaining proportions: the floor plan (see Fig. 35) and the vertical cross-section of the main nave (see Fig. 36).[7]

FIGURE 34: A pentagram in a circle in the vault of Chartres Cathedral

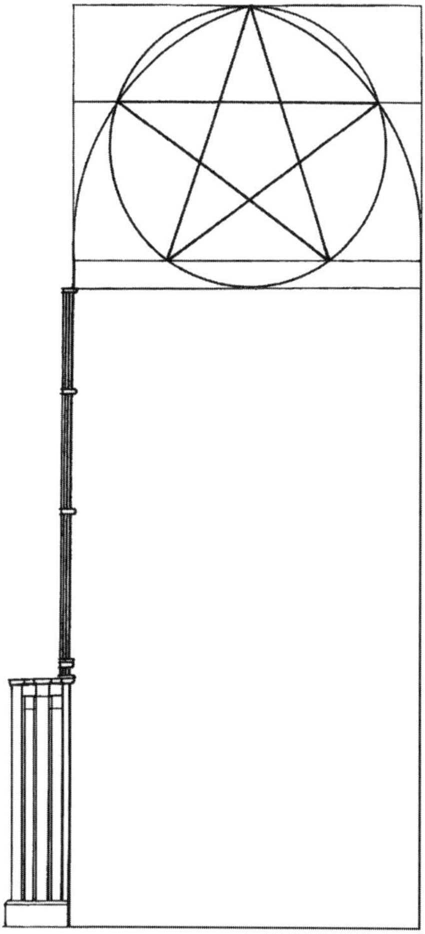

[7]The diagrams can also be found in Charles Hummel, *Pythagoras und die Meister von Chartres* (Bern: Kreis der Freunde um Hans Kayser, 1998). For fig. 34 (attributed to René A. Jeangros of Bern), see p. 18; for figs. 35 and 36, see p. 17. The source for fig. 35, in Hummel on p. 16, stems from the many other texts in which this well-known drawing also features.

FIGURE 35: The triple triangle in the floor plan of Chartres Cathedral

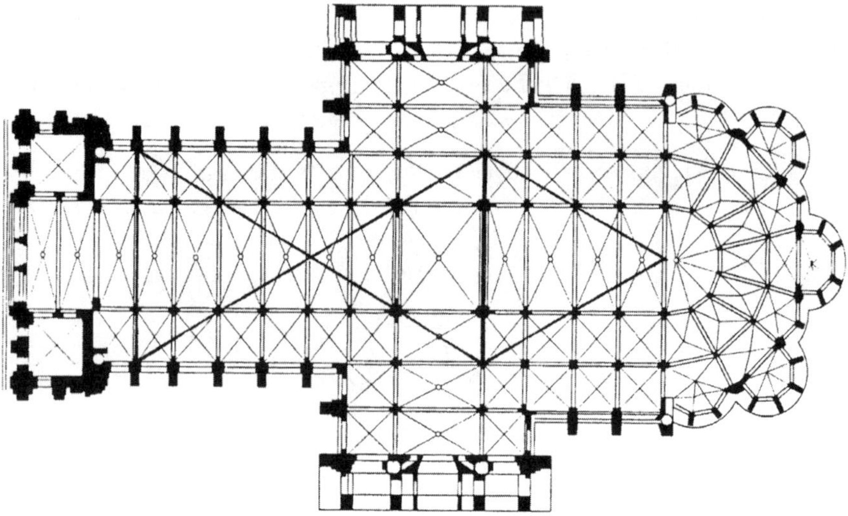

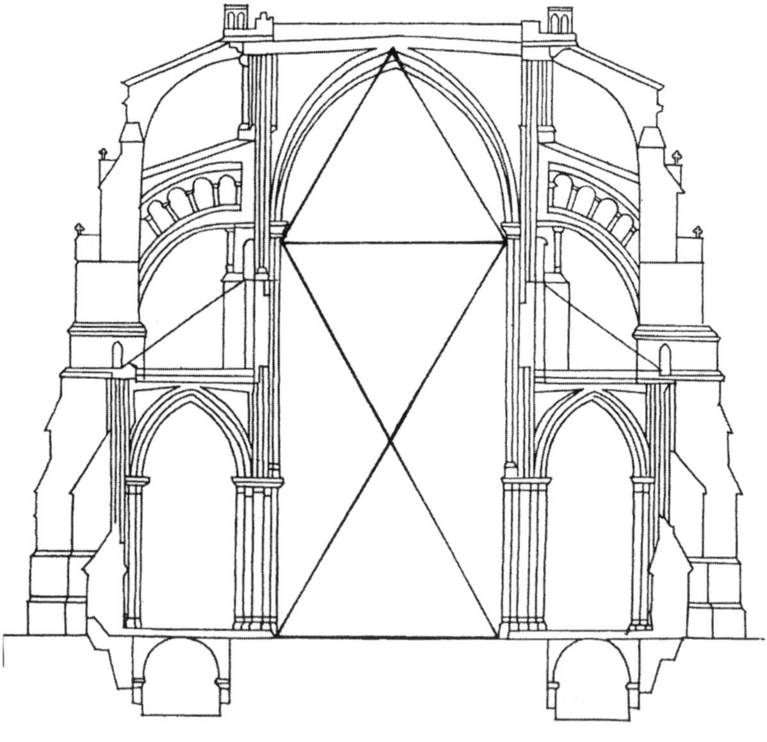

FIGURE 36: The triple triangle in the cross section of Chartres Cathedral

Nor is this thinking restricted to the masters of Chartres and Villard de Honnecourt. To give an example of a different kind, an extant medieval builder's sketch for the cathedral of Milan shows how all junctures align with a set of equilateral triangles nested inside one another at equal distances:

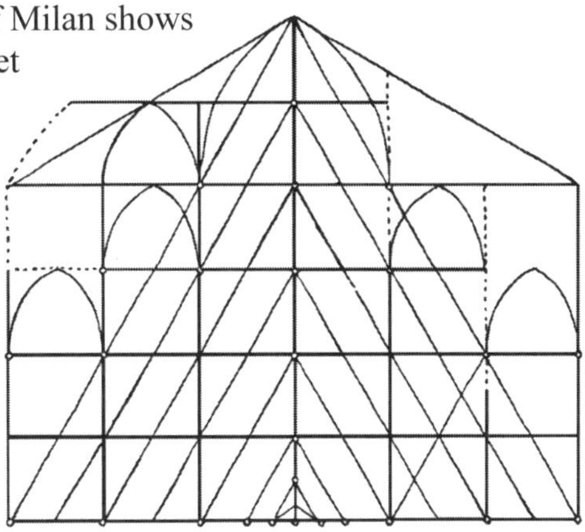

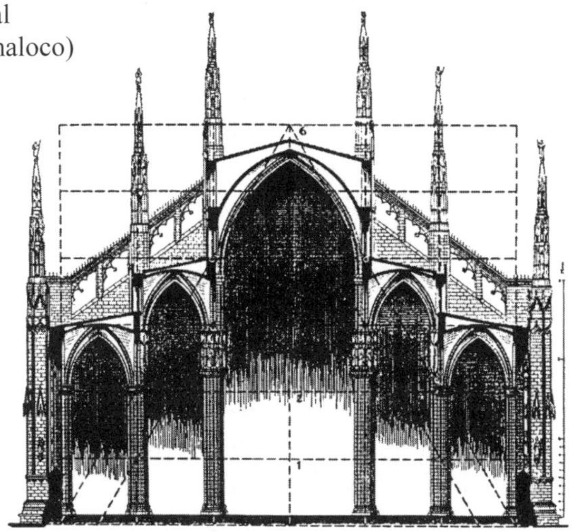

FIGURE 37: Nested triangles in the cross section of Milan Cathedral (after a 1391 sketch by G. Stornaloco)

Throughout the history of art, harmonic proportions seem to have been considered desirable also in paintings. In 1843, the Egyptologist Richard Lepsius participated in a major Prussian expedition to Memphis, the Old capital of the 1st and 2nd dynasties. In Sakkara, a section of Memphis's necropolis, he discovered a half-finished mural (dated to ca. 2650 B.C.E.) that displays an intricate web for proportional sizing. Analyzing the lines in relation to the figures, Lepsius concluded that the fundamental value the artist used was the length of a figure's foot. The size of a given foot would determine that of all other limbs and bodily features, in proportions that

Lepsius measured as $1, \frac{4}{3}, \frac{2}{5}, \frac{4}{5}, \frac{1}{3}, \frac{5}{6}, \frac{2}{9}$, etc.[8] These ratios correspond to the basic intervals (1 : 2 = octave, 4 : 3 = fourth, 2 : 5 = major third [+ 1 octave], 1 : 3 = fifth [+ 1 octave], 5 : 6 = minor third, 2 : 9 = major second [+ 2 octaves]).

Around 450 B.C.E., the greatest Greek sculptor besides Phidias, Polykleitos of Sikyon, published a theoretical work called *Canon* in which he stated the desirable mathematical relationship of various parts of the body to the whole: the proportion of finger to palm, palm to elbow, elbow to arm, etc. The proportion of the head to the body, he claimed, should be 1 : 7. While his treatise itself is lost, a statue of his by the name of *Doryphoros* (the spear bearer) survives and is generally considered to have been created in manifest support of his theories.

Vitruvius, too, devoted the initial chapter in Book III of his *Ten Books of Architecture* to a discussion of the proportions of the human body, suggesting that as the sections of a temple should relate to the whole building, so should the limbs of the human body relate to the entire human figure. A systematic development of this thought, and especially its application to character types rather than assumed norms, did not happen until many centuries later. During the Renaissance, almost all great painters spent time with the study of harmonious proportions in general and those of the human body in particular. Leonardo da Vinci's illustrations in his Venice edition of Vitruvius, in which he fitted the human figure to a square, a triangle, and a circle, are well-known examples of the norm-based variety.

More influential were the notes Dürer collected during the last years of his life. Among general treatises concerning the measurements of two- or three-dimensional figures and specific applications to the fortifications of cities, he compiled a work on human proportion, a 266-page, four-volume study illustrated with innumerable woodcuts which delves far deeper than does Leonardo famous diagram of the Vitruvian man.[9] Although Dürer never mentions music or the fact that his numbers relate to those defining consonant intervals, he is clearly addressing precisely that issue. Taking as his fundamental value the height of a human head and fitting it to bodies that measure seven, eight, nine, or ten times the head's size, he constructs a variety of character types whose other body parts he then designs in corresponding ratios.

[8]See Richard Lepsius, *Denkmäler aus Ägypten und Äthiopien: nach den Zeichnungen der 1842-1845 ausgeführten wissenschaftlichen Expedition* (Berlin: Nicolai, 1849-1913), appendix II, folio I.

[9]Albrecht Dürer, *Hierin sind begriffen vier Bücher von menschlicher Proportion* (Dietikon-Zurich: Verlag Bibliophile Drucke von Josef Stocker, 1969); facsimile reprint after the first edition (Nürnberg, 1528).

Kepler's Poetry and the Meaning of Death

The operatic protagonist never comes closer to a true understanding of how his own being partakes of universal harmony than in his musings about death. Hindemith captures the growing insight in a three-step sequence. It begins with Kepler's lament over his first wife's and little son's death toward the end of act I, is taken up on his sickbed in Regensburg with his acknowledgment that humans achieve harmony only in death, and finds its ultimate expression in the epitaph the historical Kepler composed for his tombstone, on which the protagonist reflects as he is dying. The three passages, linked by a shared melodic quotation, are thus framed by two literary products from the historical Kepler's hand—a poem and a poetic epitaph. I want to show that, while Kepler's public quest for universal harmony provides the opera's title and subject matter, its poetic core is the scientist's private poem of mourning.

Kepler's lament was occasioned by a concrete biographical incidence. Johannes and Barbara Kepler had five children. The first two died shortly after birth; the three others, a daughter and two sons, fell ill with the plague when it ravaged Prague in 1611, and six-year-old Friedrich succumbed to it. Barbara Kepler never recovered from the loss and passed away five months later. Kepler expressed his despair in an exquisite poem, which transcends personal grief with reflections about human life and death in general:

Ach leiblich Aug, du schwach Gemächt, Dein Sehen ist nur Spiegelfecht In diesen finstern Auen Du wirst, wenn hie dein Schein erbleicht, Von Angesicht zu Angesicht Das ewig Licht erschauen. Glaub sicherlich, Nit fürchte dich, Lass dir nicht kindlich grauen.	O human eye, so weakly made, Your seeing is a mere charade In our gloomy night. You will, when here your glow must fade, Come face to face with what you bade, behold th'eternal light. In faith be calm, Allow no qualm, Admit no childlike fright.
Ach Mensch, du lebst ein steten Tod. Zum wahren Leben Sterbensnot Tut nur den Anfang bringen. Auf einmal wirst du wie ein Korn Zum ewigen Leben neu geborn, Durch Christus mag's gelingen. Wünsch dir kein Weil! Durch Sterben eil, Zum Leben durchzudringen.	O man, you live a constant death. The final woe for true life's breath Can only clear the trail. Then suddenly just like a corn to eternal life you'll be reborn; Through Christ may this prevail. Seek no delay! Rush through death's fray Forth into life to sail.[10]

[10]Paul Hindemith, *Die Harmonie der Welt. Textbuch*, p. 14; translation mine.

Several features may strike the readers of this poem. In terms of its layout, the two stanzas display an intriguing ambiguity. To the *eye*, the rhymes suggest a design in nine lines grouped as 3 x 3 (in the pattern a a b, c c b, d d b). The first two sets of paired lines are iambic tetrameters; the refrain-building third lines are shortened by one foot but have feminine endings.[11] The pattern in 3 x 3 and the fact that the three groups are united by their shared final rhyme suggests a numerical symbolism that will recur in the cosmological praise Hindemith puts into his protagonist's mouth later in the opera. But there is more. Undercutting the simple regularity, the final set of paired lines in each stanza is only half as long as are the others (with only two stresses each in lines 7 and 8) and can easily be heard as a single line with an internal rhyme. To the *ear*, then, each stanza consists of eight lines. This second structural reality, and the 9 : 8 dichotomy it creates, has an analog in harmonic theory, describing one of the basic ratios, the whole tone.

In terms of its imagery, Kepler's poem addresses three aspects. The first six lines focus on the eye, more specifically the "bodily eye" (in contrast to the mind's eye): modern man's primary sense organ, his preferred gate of knowledge about the world. This eye, and the perception it grants, is described as greatly flawed and unreliable. The "gloomy pastures" of our earthly realm are insufficiently lit to permit full sight of, much less insight into, truth. The eye, and the humans relating through it to the world, are engaged in "Spiegelfechterei" (literally: fencing in front of a mirror, hence with one's own image; often translated as shadow-boxing); they lack awareness of things outside their own imagination and see only what they expect to find. Everything is illumined by the eye's own glow, the beholder's subjective light. Furthermore, as a homonym of glow or shine, the German word "Schein" also denotes semblance or sham; Kepler thus stresses once again the spurious nature of anything we can know via this sense organ. Only when the eye's glow fades can humans see the eternal light, which they will then encounter face to face—a direct substitute for the previously alleged facing of the reflected self in the mirror. Finally, Schein is the name of the composer among whose chorale tunes Hindemith chose one that might serve him to set Kepler's poem.

The second stanza's first six lines address life, in a manner corresponding to Kepler's assessment of the bodily eye. Bodily life is semblance and sham; looked at closely, it is nothing but a protracted dying process. The last moment in this process—the one humans in their delusion perceive as the

[11]Note that the pattern I describe, ˘ ´ ˘ ´ ˘ ´ ˘ ´ | ˘ ´ ˘ ´ ˘ ´ ˘ ´ | ˘ ´ ˘ ´ ˘ ´ ˘, applies fully only in the German original. My English translation reproduces the number of feet (4, 4, 3), but I am afraid I have not been able to render the rhymed feminine endings.

sole and sudden manifestation of dying—is in reality a new and more valuable beginning. Just as only the fading of the eye's glow allows an end to all make-believe and permits access to eternal light, so only the extinction of the flame of mortal life opens the doors toward true life. As with the seed that, Kepler writes recalling Matthew 13:8, must fall into the ground and die in order to bring fruit, so humans have to die in order to be born to new life.

The third aspect manifests in both stanzas' final lines. In a further analog—both switch from a rueful affirmative to an encouraging imperative—the lyrical voice admonishes us neither to fear nor to postpone the inevitable, and thereby takes the step from regret to hope. The confidence expressed in the two final groups regards not only the recently deceased loved ones, but also the mourners' own future at the other side of the gate of death. The poem as a whole expresses the conviction that earthly life is but a preliminary to true life. Kepler does not regard the valley of tears as a testing ground for the soul's nobility or obedience, but rather as the dim foyer of eternal bliss, where one should seek to dally as little as possible.

I wish to return here to the claimed analogy between the poem and the opera's layout, mentioned already in part 2. Just as the first $\frac{2}{3}$ of the two poetic stanzas each present a contrast of external and internal view, so do acts I : II and III : IV. The dualism captured in the final $\frac{1}{3}$ of both stanzas, with its turn toward hope for bliss in a life beyond the grave, corresponds to the similar dualism *within* act V.

Hindemith emphasizes the poem by means of the repeated recurrence of its musical correlate, and thereby highlights an angle that is crucial for an interpretation of the operatic plot. If his Kepler never believed that spurious earthly existence provides the precondition for a realization of ideal harmony in individual human life, then readings of the operatic plot that propose a causal link between mental goals and biography clearly miss the point. A man determined to prove that God's world resonates with perfect proportions and ratios has every right to wish that his private life might be shielded from adversities and strife; the appearance of political, economic, or other difficulties in no way proves that his quest has been misguided or inappropriate. Not only was the historical Kepler unconcerned with thoughts about whether his scientific and theological endeavors might further or hinder his private happiness; I do not see Hindemith the librettist or Hindemith the composer presupposing such a link either.[12] The protagonist clearly regrets the political

[12]Winrich Hopp seems to suggest just this in two recent essays: the *Hindemith Yearbook* article "Von der harmonikalen Fehlgründung lebenspraktischen Glücks" mentioned earlier and (a shorter version of the same argument) "Harmonikaler Wohlklang und lebenspraktischer Missklang," in *Hindemith-Forum* 6/2002: 3.

strife and personal hardship that permeate his life; but the text contains no indication that Hindemith interprets any of the afflictions as an outcome or a by-product of Johannes Kepler's professional and spiritual quest. The operatic Kepler grieves when he understands too late that instead of enjoying human harmony where it has been offered all along, at his hearth, he has sought to champion it among his official contacts. But this search in all the wrong places is a fallacy typical of *homo professionalis* in general and not limited to Kepler.

In the dramatic plot, ideas related to those expressed in the poem of mourning recur on several occasions, the most salient being the protagonist's words on his deathbed: "The great harmony ... it is death. Dying is necessary to bring it about, in life it has no stead," and, after the Regensburg pastor's last-minute nagging, "Disharmony is yet Death's song of temptation. So I hope for Heaven's harmony."

Another facet of Kepler's attitude to death—and life—is expressed in the second poetic quotation that Hindemith integrates into his libretto, the epitaph: "Mensus eram coelos nunc terrae metior umbras. Mens coelestis erat corporis umbra iacet" [I measured the skies, now I measure Earth's shadows. The mind belonged to heaven, the body's shadow lies here]. The contrast is, explicitly, between sky/heaven and shadow, implicitly, between sky/heaven and earth—the realm of eternal light and truth against that of short-time glow and shadows, the past tense of a mortal life coming to an end against the present of the moment when the dead transcends life's limitations. Kepler's body ties him to earth, shadow, and gloom; his mind sought more eternal spheres in its dual attempt at explaining the eternal laws of the cosmos and praising in them the Creator's supreme design. In the musically achieved synopsis of lament and epitaph, the diametric contrast of mortal versus true, eternal life appears as if bridged. The three-part sequence reads: bodily life / mental-spiritual life / eternal life.

Hindemith clearly endorsed this view of the value of Kepler's quest. In the center of the cosmic transfiguration that concludes the opera, he has a voice from above (the chorus embodying the Milky Way) intone:

> The energy of suns can be split for even more killing than before, or perhaps for spurring on a gentle world. But either way you will not receive an answer to the great Why. An example for us shall be the man Kepler: he sought to assimilate his life to harmony, to draw his lifer closer to it, thus to better himself, thus to raise himself up; a creature cannot attain more. [...] Nothing would have spared them, who were human, the error. But what their humble spirit perceived—dreaming, surmising, believing, praying—and their readiness to serve it: that lifted them far above the way of humans.

The quotation of the (German) poem of mourning is balanced with three Keplerian passages in Latin. Beside the above-mentioned epitaph, Hindemith excerpts from a famous prayer inserted at the end of *Harmonices mundi* as well as, later in the opera, from the passage that follows it. In the "moon rondo" of act III, the protagonist sings words of thanksgiving. The English translation of the lines Hindemith uses reads: "I thank Thee, Creator Lord, because Thou hast made me delight in Thy handiwork, and I have exulted in the works of Thy hands. I have made manifest the glory of Thy works to men who will read these demonstrations, as much as the deficiency of my mind has been able to grasp of its infinity."[13] In these words, the astronomer-mathematician shows himself again wholly devoted to his true professional aims—aims that he, in contrast to his contemporaries and scientific heirs, never tired to emphasize: through his scientific insights to reveal God's glory. In act V, the chorus of heavenly voices, welcoming Kepler on the other side of death, chides him for quarreling with the past and reminds him of "what you sang from the very beginning" by quoting the continuation of the same prayer: "Promptus mihi fuit animus ad emendatissime philosophandum" [Ready was my spirit to philosophize in the most faultless way].

Edifices of Devotion in Hindemith's *Die Harmonie der Welt*

This revelation of God's glory, of the truth and beauty of His Creation (as manifested paradigmatically in the cosmos), is both center and crowning final climax of Hindemith's musical portrayal of Kepler's story. He places his protagonist's extensive praise of cosmic and universal harmony roughly at the opera's midpoint. As briefly mentioned, the hymn, inserted as an atypical episode into the moon rondo, combines the supreme complexity and order of a fugue with the fundamental immutability of a passacaglia. (One might read the form as a metaphor for the coexistence of planets and stars in our night sky.) For the ultimate human participation in supreme harmony, the cosmic transfiguration of the principal *dramatis personae* at the opera's conclusion, Hindemith separates and expands the two aspects of the hybrid. The resulting four polyphonic segments—a fugue and first passacaglia on the one hand, a second, very different hybrid and a second passacaglia on the other—are linked by a recitative. It gives room to the final loving encounter between Susanna and Kepler, which had been missed on earth.[14]

[13]*The Harmony of the World*, p. 491 (Book V, end of chapter IX).

[14]Fugue [28-28 D]: *molto largo*, ♩ = ca. 80; passacaglia 1 [28 D-N]: *tranquillo*; ♩. = ca. 69; hybrid 2 [28 O-P]: *lento misterioso*; ♩ = 108; pas. 2 [28 P-Cc]: *moderato pesante*; 𝅗𝅥 = 92-100.

On all levels of these forms, Hindemith engages in a numerical play that proves to be based on combinations of NINE and FIVE. In the tradition of Western art, the figure NINE is much more rarely used than other numbers and thus open to new associations. It can be understood as related to THREE—be it that the larger number appears as the power of the smaller, be it that the two add up to the emblem of completion (3 x 3 = 9, 3 + 9 = 12). In Pythagorean music theory, 9 denotes the perfect tone—the only one among the three whole-tone intervals resulting from the partials and multiples considered fully "in tune."[15] Intriguingly, the musical tradition also knows a close relationship between 9 and 3: tripling the vibration of a fundamental tone twice (Hz x 3 x 3) results in the whole tone above the second upper octave; tripling the length of the fundamental's string twice gives us the whole tone below the second lower octave. Finally, in the context of an exploration of Kepler's thoughts and aspirations, 9 is also reminiscent of the line scheme in the poem of mourning whose content and structure play such a crucial role in the opera, where 3 manifests in the number of rhymed refrains in each stanza as well as in the three themes the poem explores. In the cosmic transfiguration, Hindemith seems to play with all these associations, emphasizing them in repeated couplings of 9 with 3 and depicting their precarious stability by means of conspicuous deviations.

The number FIVE was perceived already by the Pythagoreans as endowed with irrational power. As I have shown in an earlier chapter, it is connected in this opera with the more openly irrational characters, mother Kepler and Emperor Rudolf. In the context of the cosmic transfiguration, 5 is explored in an expanded aspect: the human striving to enlarge the mental faculties and gain a wider, no longer earth-bound understanding. The numerical symbol in its widest meaning thus encapsulates the desire expressed by Emperor Rudolf to gain access to "the indefinable sphere of starbright, slumbering wisdom"—to esoteric knowledge as the Pythagoreans defined it.

Hindemith's deliberate use of numerical symbolism can be demonstrated in all hymnic passages. There are altogether 5 polyphonic forms built from the same fugal subject and devoted to the praise of cosmic harmony: a hymn in act III and the four above-mentioned polyphonic forms derived from it in act V. The cosmic transfiguration in itself, which suggests a spiritual contrast to the mundane action in the preceding four and a half acts, also consists of 5 segments, since a recitative is interjected between fugue + passacaglia 1 and hybrid + passacaglia 2. Finally, the number of subject entries in all polyphonic forms plays with the figures 9 and 3.

[15]On the preference of 9 : 8 over 7 : 8 and 9 : 10, please refer back to part I of this study.

The hymn in the "moon rondo" and the four passages in the hymnic transfiguration of act V that build on ever new variations of the same material, differ with regard to the tonal and registral allocation of their subject entries. The hymn subject's twelve entries at [18] travel twice through a sequence of 5 keys (B♭, F, A♭, F♯, F♯, D♭ || B♭, F, A♭, F♯, F♯, D♭) while remaining in the low register; the 9 fugal entries and strettos at [28] emphasize the framing tone E but touch 9 keys in all (E, D, C, E; A/G, A♭/F, E♭, B, E/G/E) and involve all instruments and registers. The theme of the first passacaglia presents a rhythmically complex and much extended variant of the hymn subject, while remaining securely moored on the opera's central tone E.

Example 24: The theme of cosmic harmony with its relevant expansion

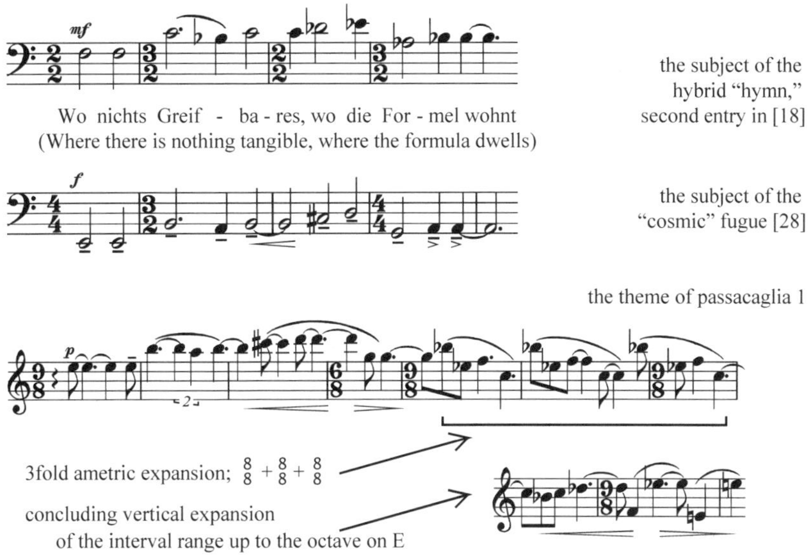

In the hymn with which Kepler praises the divinely ordered cosmos in act III, the subject is heard 12 times—9 times played by low instruments, 3 times sung by Kepler. In the fugue accompanying Kepler's transfiguration after his death, the subject manifests in 9 entries, 3 of which are strettos (i.e., group entries in stacked, substantially overlapping imitation). A cosmic chorus chides Kepler for lingering in regrets about the mistakes he made in his life. The voices invite him to affirm and proclaim what he once only divined: the truth of cosmic harmony. Sun beckons him to ascend fully and leave behind the veiled realm of Earth, where words like "futile," "important," and "in the end"—incomprehensible concepts in the cosmic realm—retain an ominous meaning.

As the cosmic allegories move further away from terrestrial concerns and limitations, their understanding expands—and so does their melodic signifier. The first passacaglia shows the *dramatis personae* in gradual transfiguration toward an idealized version of their earthly selves. Their bliss is disturbed only one more time when a final disharmony reverberates loudly enough to reach their senses. Back on earth, four assassins (among them Ulrich) have cornered Wallenstein. They charge him with cruelty, bloodshed, treason, despotism, accuse him of being "a false tone in the chord of the worlds," and stab him to death.

With this, planet earth, the realm of finite and flawed human beings, disappears from view. As it fades, Kepler (now allegorized as Earth) joins the heavenly gathering and takes his place behind Mercury and Venus in the row of planets. The allegories arriving after him, Mother Kepler as his Moon and the three outermost planets, each try to salvage for the cosmic dimension some aspect of what defined them in human life. We hear justifications of magic (Moon/mother Kepler), of contradiction as a means toward insight (Mars/Ulrich), of glory as a principle within a hierarchically perceived human order (Jupiter/Wallenstein), and even of a cynical attitude toward all life and death (Saturn/Tansur).

In the course of the first passacaglia, which is devoted to this transformation, the thematic phrase is heard 10 times: an odd "extra" disturbs an underlying regular entity of 9. These 9 entries are carried out by varying instruments, which accompany the cosmic gathering describing a gradual crescendo from ***p*** to ***ff***. Quite early in the process, the measured build-up is upset when segments of a trumpet-led statement are suddenly doubled by vocal utterances that burst out in rough *forte*. Significantly though, the voices exhibiting this unruly behavior are not those of the cosmic allegories. Instead, they belong to the four murderers, resonating from violence-ridden earth into the heavenly serenity. Startled, the cosmic impersonations turn their eyes to humankind's by now distant home planet, where they see Ulrich and his three military companions approaching Wallenstein. By the time the murder is committed, the cosmic impersonations have long turned away in sad acceptance of humans' erroneous ways. Just as this murder interrupts the incipient cosmic calm, so the redundant tenth subject entry disturbs the passacaglia theme's perfect number of 9 statements.

A very similar shape, now rhythmically simplified and restricted to two keys, underlies the two single entries and one stretto of the second hybrid form. As if to offer a compromise, the theme of the second passacaglia, once again firmly rooted in E, blends the original shape of the hymn subject with the extension developed in the earlier passacaglia and confirmed in the hybrid's *misterioso*.

EXAMPLE 25: The expanded theme of cosmic integration, with variants

the theme of passacaglia 1

the theme of the "mysterious" hybrid

the theme of passacaglia 2

[How, then, could anything that is as limited and small as we are be causally united with a formula that governs the whole cosmos?]

During the finale's next polyphonic component, marked "slow and mysterious," the planets begin to grasp the immensity of the design of which they are a part. Understanding that they are indissolubly bound as siblings of the light, they claim a new identity "as voices in God's chorale." The theme, once again restricted to the accompanying instruments, manifests in 3 statements. Twice it sounds in very soft *pizzicato* / *staccato* and in C♯, with a truncated end that fails to achieve tonal or metric closure. Its third entry, by contrast, is a stretto on A whose various imitations all break or taper off. While the text suggests that all selfish interests will now be given up, listeners witness the self-regulated demise of last attempts to propose a tonality deviating from E, the anchor of universal harmony.

The second passacaglia features precisely 9 entries. Once again the number 3 plays a secondary role, complemented with a transparent tonal symbol: the last 3 entries are varied and leave the tonic E to ascend through three consecutive perfect fifths: B, F♯, C♯. In terms of the quest for universal harmony and its practical application, this final passacaglia provides a conclusion. The chorus, now visible on stage as "the Milky Way," states that given the insignificance of individuals, a postulate like that of universal consonance can never be used for any purpose, be it good or bad. When the voices mention Kepler, who sought to assimilate his life to harmony in order to become a better person—the best that can be expected of any creature—this emphasis on an individual momentarily disturbs their glorious serenity, and the thematic phrase withdraws. At the reminder of Kepler's errors, the strings launch chains of $\frac{5}{8}$- figures that sound polymetrically against the winds. As soon as the stars of the Milky Way abandon their musings about what universal harmony may mean on earth and bask once again in their eternal "soaring in highest roundelays," the passacaglia resumes its regular phrases. Finally, all allegories (chorus and ensemble: Milky Way, Sun, Moon, and planets) praise the infinity of the universe and express the hope that their current experience may raise each of them above the narrow Self, to merge into the great harmony of the world.

When a scientist strives to describe the eternal beauty of divine Creation, or when an artist attempts to imitate in the work of his hands the perfection innate in the natural world, the results can never reach the glorious harmony reserved to the Creator God. Human beings, who in their lives inevitably generate disorder, return to order only in death. Thus it seems convincing that Kepler's poem of mourning with its meditation on death, and Hindemith's opera—most notably in those segments that address the praise of harmony explicitly also on the dramatic level—come as close as any fallible human being can to the goal of appropriately extolling the Creator. In the words of the cosmic voices: "a creature cannot attain more."

Hindemith made sure that perfect proportions pertain also in dimensions that no human sense organ can consciously ascertain. The ideal durations of the five polyphonic segments according to the composer's indications are "harmonious" in just that sense: they relate to one another in the simple ratios that correspond to musical consonances:

hymn : fugue : passacaglia : hybrid : passacaglia = 5 : 8 : 9 : 3 : 9.

If Kepler could "hear" the relative angular speeds of planetary orbits as intervals, surely we are permitted similarly to claim "hearing" the harmony in the relative durations of the corresponding musical segments. With reference to the fundamental E of both Pythagorean musical philosophy and Hindemith's Kepler opera, the praise of cosmic order in *Die Harmonie der Welt* then sounds in this glorious contour:

EXAMPLE 26: The "consonance" of the hymnic passages

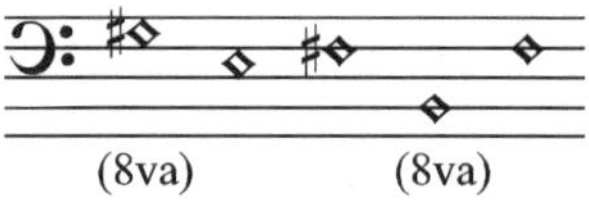

Knecht's Awakening and Death

Joseph Milek characterizes Castalia quite devastatingly as "a world that has proscribed artistic creativity, ostracized women, banned marriage, and sublimated sexuality, knows no politics, practices no religion, cultivates collegiality, nurtures anonymity, and frowns upon psychology, is unconcerned about history, untroubled by fate, and has become oblivious of time."[16] With such a catalogue, the 23rd-century stage of the World of the Mind would seem doomed.

Yet an outstanding *magister ludi* of the era, Joseph Knecht, breaks with several of these rules. During his mission at the Benedictine monastery, he is involuntarily drawn into politics; from Father Jacobus, he learns to respect the importance of history; and in conversations with Plinio Designori, he becomes concerned for Castalia's fate and conscious of the temporal dimension of its endeavor. Knecht knowingly breaks with the Order's customary reserve by speaking of his awakening experiences. When he decides that he must leave his position and the Province to become a simple teacher or tutor, he does not make up glib and acceptable excuses but justifies his conviction with the urge to heed his inner voice—an admission that seems so out of place in the Castalian atmosphere as to prompt the embarrassed President to attribute it to "an excessive sense of your own person" (p. 396).

In a world where obeying the rules of conduct and decorum grants all the smoothness that is required, Knecht's individual conscience impels him to transcend what is thought to represent the highest stage, to seek harmony beyond what his fellow Castalians believe feasible or even desirable, and to find death. Last but by no means least, Knecht violates the ban against artistic creativity by engaging in what is most frowned upon: "writing poetry was regarded as the most impossible, ridiculous, and prohibited of conceivable acts" (p. 106). Knecht's biographical path toward an integration of the World of the Mind with the natural world, for which his startling but utterly logical death both lays the basis and establishes a symbol, and his mental trajectory toward the same goal, which (given the place it occupies in Hesse's work) climaxes in the expression of his inner struggle in verse, constitute parallel closures to the two strands of the double helix described earlier.

Hesse could no doubt have concluded *The Glass Bead Game* in a number of ways that might have been less anticlimactic for the reader. But his 25th-century fictional biographer, I contend, could not. For this died-in-the-wool Castalian, Knecht's defection from the highest position the elite

[16]Joseph Milek, *Hermann Hesse: Life and Mind* (Berkeley: University of California Press, 1978), p. 310.

system had to offer—his abandonment of what amounts to an equivalent of the papal see, as it were—threatens the foundation on which he and his peers have built their entire lives. The chronicler may or may not know how the remainder of Knecht's life developed; he is almost certainly not interested, if for no other reason than that he cannot imagine wanting to explore what such a life "in the world out there" might be like and what it might, God forbid, have to offer one who has tasted the richest fruits of Castalian sophistication. At the same time, his respect for the revered *magister ludi* does not allow him to present Knecht's existence in the world as a simple failure, and then gloat about this predictable outcome.

The solution he finds is perfect—if not for us, then for him and his likes. Knecht's accidental and almost (but not quite) incomprehensible death in the cold waters of the mountain lake can safely be assumed to fill everyone with sadness. The distress felt is mitigated by the absence both of prolonged suffering and of external malice. Should anyone secretly judge that none of this would have happened had Knecht remained in office, where he would have served the community rather than his private whim—well, such thoughts would be the thinker's responsibility and should not be blamed on the biographer.

While the biographer may thus have his own motives for wishing to make short shrift of the private afterlife of an exemplary Castalian figure, Hesse the mastermind presumably had no reason to resent Knecht's decision to transcend his world in search for a higher synthesis. But the afterlife that interests Hesse is that of Knecht's thoughts—particularly those captured in his clandestinely written poems. These he has his chronicler append to the work, wedged between "Life" and "lives," as if to stress their timelessness. The poems resonate in the air just as readers are pondering what could or should have happened to a *magister ludi* turned secular citizen. They suggest an answer to the quest for harmonization—and this answer does not depend on the continued physical existence of the man who wrote them.

Knecht's Poems: Synthesis of a Discursive Glass Bead Game

The discursive glass bead game, whose first four movements I analyzed much earlier in this study, comes closest to the aspired synthesis in the two components that result from circumstances jarring with Castalian expectations. One is a conversation in which the *magister ludi* discusses the state of Province and Order with a man from the outside world; the other, a circular in which he confronts the Castalian Directorate with a mix of premonitions, analyses, and the startling request to be released from his office. Thus it seems consistent that the final step in the harmonization process should happen in a medium that Castalians would have frowned upon even more: in the form of rhymed pentameters.

The thirteen "Poems of Knecht's Student Years," with which the chronicler opens the section of his protagonist's posthumous writings, go beyond the juxtaposition of two worlds that dominates the first and third movements. Instead, the poems expand on the dialectic of perceived distance and desired harmony between these two worlds, notions that inform the thematic material of the second and fourth movements, and lead this dialectic toward a solution. As was to be expected, the approximation and ultimate harmonization is achieved by the glass bead game players.

An echo of the sensitivities the two worlds have long felt toward one another is heard one last time at the beginning of the series, in poems 2-5. These four poems, which on the surface appear as independent entities in the larger series, are actually arranged in two pairs. These pairs juxtapose the contrasting needs and ambitions each world feels as well as the respective fear of being crushed or smothered by the other.

In poem no. 2, entitled "A Compromise," young Knecht accuses "die ewig Unentwegten und Naiven" [the eternally busy and naive] of being blinded by their desire for a world that is simple, safe, and therefore flat. As he tells it, these people not only decry any doubts others might have, but even go so far as to deny that there could or need be any depth. Their choice regarding how much of the world they care to acknowledge has audible repercussions: by "cancelling one dimension," they upset the poetic logic, provoking a void where one feels there should have been an additional line, and thus causing the sonnet's first tercet to appear truncated. Conversely in poem no. 3, "But Secretly We Thirst . . .," the playful dream-dancers discover that they crave reality, even long for a taste of blood and barbarism.

In poem no. 4, "Alphabets," a savage is so frightened by the intolerable incomprehensibility of a piece of script that he subjects the threatening ciphers to a ritual burning, hoping thereby to destroy anything they might stand for.[18] As a counterpart in poem no. 5, "On Reading an Old Philosopher," the erudite have the daunting experience of seeing the written essence of their centuries-old cultural heritage lose its meaning-granting context, in front of their very eyes and without any discernible aggressive influence

[18]The Winstons' rendering of Hesse's title "Buchstaben" as "Alphabets" successfully avoids the ambiguity that the literal translation—"Letters"—would create in English. But it does so at the price of evoking not the single ciphers that combine to units of meaning accessible to the initiates, but the more generic notational systems of various languages in conventional ordering—which are mere tools and do not in themselves signify any object or concept.

from outside. Turning into mere arabesque and losing all power of expression, the script comes to resemble a musical notation from which all "clefs and sharps" have suddenly been erased.

The principal theme in this fifth "movement" of the discursive glass bead game is thus the experience of impermanence, which applies to all worlds, be they built on mind or nature, on intellectual pursuits and asceticism or on the rich though often chaotic textures of emotions and cravings. The secondary theme, juxtaposed with the notion of transience, is that of cipher notation. This kind of notation is presented as the only medium potentially capable of counteracting the pervasive force of corrosion, owing to the fact that it creates no products that are temporally and spatially defined and thereby destructible, but only their codes. As a script invented to capture signifiers for contextual units, it preserves a symbolic essence. At any time (be it hours or eons after its creation), such an essence must be brought to new life in an act of devoted interpretation and, perhaps, "performance."

The idealized parallelism suggested in this image—of music on the one hand, the glass bead game on the other—confirms a correspondence that has been resonating throughout Hesse's compound work. By hinting that the concrete game as it has developed within the Province of Castalia must eventually perish along with its players, while music will not, the student Josef Knecht declares his most beloved art as even closer to the spirit that informs his world than the game that serves as its acknowledged symbol.

As was the case in the preceding four movements of what I call a discursive glass bead game, the two contrasted themes are complemented with three motifs. In motif 1, the poet ponders the ambivalence of the human potential, which seems so promising but proves so limited. With motif 2, he confirms "service"—thematically highlighted also in the two preceding movements—as the attitude that befits all humans, but especially those who live directed by concerns of the mind. With motif 3, "serenity," he takes up what functioned as a remedial concept in the Knecht-Designori conversation (see motif 3 in discursive movement 2) as well as, indirectly, its cause, meditation (motif 3 in discursive movement 1).

The recognition to what degree human potential remains hemmed in by impermanence is succinctly captured in the third quatrain of the initial poem, "Lament," where the two notions are paired in thoughts about humans as instruments of divine whim:

> What God would make of us remains unknown
> He plays; we are the clay to his desire.
> Plastic and mute, we neither laugh nor groan;
> He kneads, but never gives us to the fire.

When pondering the human dependence on a divine potter, the Waldzell student Joseph Knecht adds a nuance to the account of religion in Castalia, which in the chronicler's presentation has remained rather pale. In the reflections of poem no. 11, Knecht seems to regret that the view Thomas Aquinas held of man's relationship to Mind/Spirit—"Von Gott her zu Gott hin der Mensch gestaltet" [the human being molded as coming from God and intended to return to God]—is no longer pertinent. Similarly, in thoughts (inspired by a Bach toccata) about the world's creation, Knecht describes the mighty urge that "shines" everything into being as "gottwärts rückbezogen" [defined by its retroactive relationship to God]. Admittedly, this God is a generic Creator and not the first person of the Christian Trinity. Nonetheless, even the Castalian's relationship to this divine potter is once conceived as personal when, in poem no. 7, Knecht speaks of "Vater Geist" [Father Spirit-Mind]. In the later poems, however, the anthropomorphic image gives way, in no. 12, to that of a "Weltgeist" [World Mind/Cosmic Spirit] and finally, to the notion of "the sacred center" ("die heilige Mitte," which the Winstons translate as "Cosmic Soul").

In poem no. 3, "But Secretly We Thirst . . .," Knecht contrasts his image of a Creator-prescribed human potential with thoughts about what happens when people take the definition of their options into their own hands. The dream-dancers who devote their lives to gracious play—a transparent image of less-than-ideal glass bead game players—end up thirsting for aspects for which their life plan does not provide; even suffering and death are recognized here as part of what defines the full human potential. In poem no. 10, "Soap Bubbles," Knecht believes to understand that all human achievements have ultimately the same, utterly transient value. Finally in no. 12, "Stages," he grasps that the key to the full realization of the human potential lies in remaining adaptable, forever ready to answer new calls.

The motif of "service" adds a further dimension to the duties the World of the Mind owes the world-at-large—duties the mature *magister ludi* outlines in his famous circular. In poem no. 9, which Hesse entitles "Dienst" [service],[19] Knecht observes that what is required in this time of decline is

[19]The translators, reading "Dienst" as short for "Gottesdienst," render the title as "Worship," thereby narrowing its scope. They also avoid the phrase that picks up the larger meaning of the term in the fourth stanza, "unser ist das Amt." As a word that can generally imply either a service to be rendered by a person thus predestined by office, or the office itself (cf. "civil service" etc.), "Amt" confirms the general rather than the religious meaning of "Dienst." Moreover, where "the sacred line of the sons of the gods has long been extinguished," actual worship is probably not even called for. By contrast, efforts to preserve a knowledge of humankind's higher point of reference are very much in need. Employing their sophisticated means to this end is the "service" on which Castalians, in young Knecht's view, should focus.

that the memory of sacred reverence to a power beyond the human sphere—here imagined as a sun-god—be preserved "durch Zeichenspiel, durch Gleichnis und Gesang" [through cipher-based game, parable, and song]. Furthermore, in two poems addressing tomes of accumulated wisdom that prove disturbingly prone to self-generated destruction (nos. 5 and 8), Knecht concludes that the service their authors have rendered humankind is not lost together with the material substrate, for

> Man's spirit, struggling incorruptibly,
> Painfully raises beacons, death defying,
> And wins, by longing, immortality.

A disposition that is "heiter" [serene, in the sense both of tranquil and cheerful] defines the playful dream-dancers who secretly thirst for an experience of life's dark side, but also those who, reading Thomas Aquinas, believe to be entering a beautifully measured temple in which truth shines brightly. So far, not much has changed: serenity is the domain of the erudite ascetics of the World of the Mind who can afford the time to meditate regularly and thus achieve a calm detachment. What Knecht's poems add to this view is the notion of "genesen" or "gesunden" [recover, convalesce], used explicitly here with regard to a person's spiritual state.

This expansion of the concept of serenity underscores the harmonizing function of the discursive glass bead game's closing movement, since it applies to all humans and not only to the elect. In "Alphabets," the savage only reaches a state where he can "sigh, smile, and recover" after he has seen to it that

> diese Un-Welt, dieser Zaubertand,
> Dies Unverträgliche zurück ins Niegewesen
> Gesogen würde und ins Nirgendland
>
> [this anti-world, this magic tinsel,
> this irritable stuff would be sucked back
> into the never-has-been and the nowhere-land]

And in the poem "Stages," the poet recognizes that

> Even the hour of death may send
> Us speeding on to fresh and newer spaces,
> And life may summon us to newer races.

By meeting each of life's summons with the readiness for change and new beginning, every farewell will turn into a recuperation.

Musical (and Game) Notation: Antidotes to Impermanence

So much for the three motifs that qualify the leading thematic components in the poetry series. The two dialectically contrasted themes differ from those of the preceding "movements" not only in that they represent a higher level of abstraction, but also in that they are raised to a synthesis.

The principal theme, "impermanence," is prevalent in more than half of the thirteen poems. The collection opens with a lament mourning that "No permanence is ours; we are a wave,"[20] and proceeds to suggest that humans feel so desperate about this transience that they might even wish to congeal into stone—just so that they might experience at least once what it means to last without undergoing constant change. In the sixth poem, "The Last Glass Bead Game Player," we witness the moment when the extinction of an entire caste is imminent as its only surviving member drifts into senility. The ninth poem confirms the death, quite a while ago, of the last of an ancestral line of gods. Poem no. 8, "A Dream," chronicles how an avid book lover, glorying in the intellectual wealth of the wisdom accumulated in a monastery library, discovers that neither the ink on the parchment nor the relationship between a tome's title and content are reliable; everything might change at any moment, and all answers to life's questions are ultimately in the eyes of the beholder. In "Soap Bubbles" (no. 10), the book of wisdom written by a wise elder and the ambitious treatise of an avid young scholar are likened to the shimmering water globules blown by a little boy.

The secondary theme—in general: any formulaic script that overcomes material transience; in particular: musical and glass-bead-game notation—is played out in two poems entirely devoted to this topic. In "A Toccata by Bach" (poem no. 7), music seems capable of creating an entire world, with its temporal as well as its spatial dimensions. The moment of expectation before the music begins is likened to the unformed, virginal existence of the universe before Creation, a state neither dead nor wholly passive: darkness is said to "attend to its duties" ("Finsternis waltet") and primordial silence, to "gaze fixedly" ("Urschweigen starrt"). From within this state of congealed potential, the forces of creation then erupt. A shaft of light breaks from a tear in the clouds and "seizes world-depths from blind nothingness" by erecting spaces. It burrows through the night, creating by making visible what yet lacks name and purpose just as music creates by making audible what has no pragmatic connotation. The shaft of light in its creative act is divisive, generating cell growth but also deeds and wars. It startles the world, which begins

[20]Or, as a more literal translation of Hesse's "Uns ist kein Sein vergönnt. Wir sind nur Strom" would read: "We are not favored with Being. We are a mere flow."

to shine, to fall into order, and to praise life and its Creator in splendid sound. Remaining forever determined by its source in God, the generative light—heard as music—propels its way through bustling creature activities toward the "Father Spirit." It is both pleasure and sorrow; it manifests in all the arts ("language, image, song"); it is instinctual urge and spirit-mind, strife, happiness, and love.

The glass bead game, too, in its deepest truth is a form of music. In the central verse of the collection's final poem, entitled "The Glass Bead Game," Knecht speaks about a magical script of formulas in such a way that the term could apply equally to music proper (the "music of the masters" mentioned early in the poem), to cosmic consonances (the "music of the cosmos" with which the poem opens), and to the Castalian play with cultural essences evoked in the title. This cipher-based script is celebrated for enabling those who practice it to be uplifted by its mystery, and for congealing into "precise parables"—and thereby rescuing from impermanence—what is otherwise stormy and a flow without shores ("uferlos").

The verbal iconography of the unbounded flow congealed into permanence connects this closing poem with the opening one. But there is a significant difference. In "Lament," the poet expresses regret that the divine potter never decides to give any of the human forms he is forever kneading to the fire; in "The Glass Bead Game," humans themselves have found a symbolic equivalent of the kiln, in which not stages of (individual or phylogenetic) development, but their potential as it is cast into intellectual and artistic expressions, will be preserved.

This change of insight informs not only the framing poems. While many of the various images evoked in the interior verses seem to mourn impermanence as a condition of life that is much to be regretted, young Knecht's attitude vis-à-vis this inevitability undergoes a development, so that towards the end of his series he sounds a note that is clearly different from those heard toward the beginning. Thus the eleventh poem is a meditation about the extent to which each era, largely because of the impermanence of many of its traits, is invariably idealized by later generations. Pondering the relationship between all that which habitually passes into oblivion and that which remains, the poet shifts his accent from sorrow over the loss to an acknowledgment of life's wisdom. Transience, he observes, pertains above all to personal doubts and pains, whereas the signifiers that the (individual or collective) mind distills from these personal experiences stand a good chance to remain.

Similarly, the much-quoted poem "Stages" embraces the notion that constant change is not invariably regrettable but in many regards necessary and beneficial. Hesse's young Knecht thus corrects his perception that humans endure impermanence in an exclusively passive way. Instead he declares it an active choice—a decision to let go of any achievement and leave any platform reached whenever a new task summons, to be ready for transformation in order to realize ever new dimensions of one's potential.

Interestingly, this poem—the second-to-last in the arrangement chosen by the fictional Castalian chronicler, who evidently felt the need to close with a poem about the glass bead game—is effectively the last to have been composed, both literally by Hesse and allegedly by young Knecht. As Volker Michel's excellent volume of *Materialien* informs its readers, Hesse wrote the other twelve poems between December 1932 and November 1937;[21] "Stages" emerged as a kind of afterthought in May 1941, after both the "Life" and the "lives" had been completed. And while the first twelve poems are reported in the story as stemming from the years of Knecht's debating joust with Plinio Designori during his time at the Waldzell school (see pp. 105-106), we learn on occasion of the *magister ludi*'s leave-taking from his Castalian friend Tegularius, when he remembers his poem about the "stages" of life, that Knecht had written this one "during his stay in the College of Far Eastern Studies" (p. 373), hence at a much later stage in his own development.

In the course of his early maturation process as it appears traced in his poetry, Knecht understands that any change concerns only the surface, never the essence: Mind is immortal, even if its products are not: "intimations of the true life never died" (poem no. 9, line 13). Every human attempt to give expression to Mind counteracts impermanence (11/34-36):

> For in us too is part of that Eternal Mind
> Which through the aeons calls to brothers of its kind
> Both you and I will pass, but it survives.

In this process of preservation, the notions born from human ambition have no say: the illusions born "out of the Maya-foam of the universe"—equally transient whether they be a child's soap bubbles or a scholar's treatises—have value in that "in each of them the Light of Eternity sees its reflection, and burns more joyfully" (10/14-17).

[21]Viewed in the order of their creation, the poems proceed from the glass bead game players' "secret thirst" for the less intellectual aspects of life to the last player's gliding into senility. The dates of composition are: 12/1932 "But Secretly We Thirst . . ." (no. 3), 8/1933 "The Glass Bead Game" (no. 13), 1/1934 "Lament" (no. 1), 2/1935 "Alphabets" (no. 4), 4/1935 "Worship" ["Service"] (no. 9), 5/1935 "A Toccata by Bach" (no. 7), 6/1935 "After Dipping into the *Summa contra Gentiles*" (no. 11), 7/1936 "A Dream" (no. 8), 11/1936 "A Compromise" (no. 2), 11/1936 "On Reading an Old Philosopher" (no. 5), 1/1937 "Soap Bubbles" (no. 10), 11/1937 "The Last Glass Bead Game Player" (no. 6), 5/1941 "Stages" (no. 12).

The greatest resistence to the destructive forces of impermanence is offered where reliance on material manifestation is minimal and the degree of abstraction highest. This applies to two realms: music and the glass bead game. In both, players take their ever-new, ever-old information from cipher-based notation, which suggests but never claims fully to capture what is intended but must be re-created in each new moment of time's flow.

Coda

Music in its ideal realization—as Hesse found it embodied in Bach's compositions—procures a synthesis between the worlds of Mind and nature. Rather than treating them as initial opposites that are to be coaxed into greater proximity, it addresses them jointly as parts of a larger unity. Thus music is, for Joseph Knecht, the harmonizing force in both meanings of this word. The concluding poem begins with a remembrance of the power of music. The wording of Hesse's initial line seems intended to evoke Boethius and Kepler, insofar the heavenly constellations of *musica mundana* appear next to the best of *musica instrumentalis*.

> Musik des Weltalls und Musik der Meister
> Sind wir bereit in Ehrfurcht anzuhören,
> Zu reiner Feier die verehrten Geister
> Begnadeter Zeiten zu beschwören.
>
> [Music of the cosmos and music of the masters
> We are ready to hear in a spirit of awe,
> To conjure up the revered spirits of blessed times
> For pure celebration.]

Music is here not so much what is actually realized in sound, but rather all that is consonant: a symbol of the order underlying every seemingly unformed manifestation of earthly life. Humans living under the direction of Mind serve the various realizations of this music. In this service they find their highest calling, their true destiny, which is entirely free of anxiety. For nobody who has become part of this circle can fall in any direction other than toward the sacred center, to which all paths lead: to the immortal, forever unchangeable Spirit-Mind itself.

Conclusion: Musical Order in the World

For almost exactly two thousand years, the followers of the Greek and Chinese wisdom traditions—from Pythagoras (ca. 570-496 B.C.E.) to Kepler (1571-1630 A.D.) and from Confucius (ca. 551-479 B.C.E.) to the end of the Ming dynasty (1644 C.E.)—believed that the world as a whole and each event or phenomenon in it carry meaning. The meaning is understood as spiritual in its essence but capable of manifesting itself in such a way as to shape material substances. Humans must seek to grasp the intended and underlying order and adjust their individual life decisions to it.

In the mid-20th century, Paul Hindemith and Hermann Hesse, living through a period of particular chaos and distress, composed works centering on protagonists devoted to the quest for order and harmony. Hindemith chose an historical astronomer, philosopher-theologian, and polymath, Johannes Kepler, whose research for his treatise *Harmonices mundi* constitutes a late pinnacle within the Pythagorean tradition. The laws he discovered mark a turning-point of scientific thinking that ushered in many of the developments on which the understanding of our time is still based. Hesse invented Joseph Knecht, an erudite musician and glass bead game player of a future age who is deeply influenced by the Chinese wisdom tradition. His insight into the urgently needed harmonization of the worlds of mind and nature appears as a turning-point in the era following the alleged "intellectual reformation" of the late 20th century.

While addressing aspects as different as the cosmos and its complex but regulated motions, the defense of individual religious conscience versus prescribed doctrine, the formula-based play with the contents and values of all cultures, and the mutual support of ascetic intellectuals and people of the world, the two creative artists and their heroes share an essential feature: they can be shown to derive their sense of order from music and to couch their attempt at re-establishing order into music.

One of the core tenets of Chinese wisdom, captured particularly in the *I Ching* [The Book of Changes], is perpetual transformation: the constant transition from one state into another. The resulting temporary constellations are not, however, believed to be brought about by external forces. Instead, they themselves *are* a play of forces. Behind the world of appearances exists an entity that *is* the totality of changes, the eternal law that manifests in all motions and transformations. This is their meaning, their *dao*.

Kepler, like all Pythagoreans before and after him, would have agreed. As demonstrated in *Harmonices mundi*, he saw a single law expressed in all aspects of nature. He regarded this law as spiritual, directly accessible to the human senses only in the form of tonal relationships. Hindemith, in many of his compositions and theories but especially in his opera *Die Harmonie der Welt*, extrapolated from Kepler's insight. Through the myriad aspects of his music, the composer reveals an underlying order in events and processes that by themselves appear lacking in any higher significance.

In a similar spirit, Joseph Knecht grasps the law of impermanence. In his world of post-Christian asceticism, the essence that undergirds and unites all transient states and is itself eternal is "spirit-mind." Besides its unspecified form—which, like the *dao*, can only be addressed with awe but cannot be defined—it also exists in the distilled state given it by the best of human minds: in the formulaic notations used by glass bead game players and musicians. Hesse builds his layered literary text on his protagonist's apprehension when he structures the philosophical of two strands in analogy with a musical composition.

During the decades following the publication of Hesse's and Hindemith's works, Western music often seemed threatened to lose the capacity of manifesting a spiritual order. The ancient tonal system based on what would later be known as partial tones, the common-period system based on well-tempered versions of triadic harmonies, and even the various modern successors built on humanly constructed artificial modes: all seemed doomed to be abandoned in favor of arbitrariness. When, moreover, neither structures nor textures seemed resilient enough to instill hope for some *dao* indestructibly surviving our despondent cacophony, artist-seekers turned to subject matter, rediscovering religion and myth—even saints and Christ himself as operatic protagonists.[1]

Hesse's Castalian chronicler would have diagnosed the final phase of the Age of the Feuilleton, and comforted his readers that this phase, too, eventually passed into oblivion. Joseph Knecht might have connected our state of music and art to his observation that ever since the 19th century, the crucial balance between spirit and emotion has been abandoned in favor of the latter. Just as he seeks to harmonize Castalians with worldly people in *The Glass Bead Game*, he would have devoted himself to remarry spirit and emotion in 21st-century art.

[1]See my extensive studies on these topics: *Saints in the Limelight: Representations of the Religious Quest on the Post-1945 Operatic Stage* (Hillsdale, N.Y.: Pendragon, 2003) and *Christus als Opernheld im späten 20. Jahrhundert* (Waldkirch, Germany: Gorz: 2005).

Appendices

I: Perspectives on the Glass Bead Game

(1: Hesse's commentary on the projected fifth life in his 1931 sketch, transl. from V. Michels, ed., *Materialien* I, pp. 314-315.)

Highest culture: the bead game in many categories, embraces music, history, space, mathematics. X is now the highest of bead players, plays the world symphony, varies it in accordance with Plato, with Bach, with Mozart, expresses very complex aspects in ten rows of beads, is fully understood by only three or four, halfway by thousands. [...] The intellectuals have ceased writing books, instead of which [they play the] bead game. They have also renounced comfort and success, and live, most contented and free of wants, for their beautiful lifelong game. Description of the game: "not easy to render evocatively, since so complicated and at any rate not yet invented."

(2: pp. 15-18, the chronicler at the outset of his introduction)

The Glass Bead Game is thus a mode of playing with the total contents and values of our culture; it plays with them as, say, in the great age of the arts a painter might have played with the colors of his palette. All the insights, noble thoughts, and works of art that the human race has produced in its creative years, all that subsequent periods of scholarly study have reduced to concepts and converted into intellectual property—on all this immense body of intellectual values the Glass Bead Game player plays like the organist on an organ. And this organ has attained an almost unimaginable perfection; its manuals and pedals range over the entire intellectual cosmos; its stops are almost beyond number. Theoretically this instrument is capable of reproducing in the Game the entire intellectual content of the universe. [... W]ithin the complicated mechanism of this giant organ, a whole universe of possibilities and combinations is available to the individual player. For even two out of a thousand stringently played games to resemble each other more than superficially is hardly possible. Even if it should so happen that two players by chance were to choose precisely the same small assortment of themes for the content of their Games, these two Games could present an entirely different appearance and run an entirely different course, depending on the qualities of mind, character, mood, and virtuosity of the players.

How far back the historian wishes to place the origins and antecedents of the Glass Bead Game is, ultimately, a matter of personal choice. For like every great idea it has no real beginning; rather, it has always been, at least the idea of it. We find it foreshadowed, as a dim anticipation and hope, in a good many earlier ages. There are hints of it in Pythagoras, for example, and then among Hellenistic Gnostic circles in the late period of classical civilization. We find it equally among the ancient Chinese, then again at the several pinnacles of Arabic-Moorish culture; and the path of its prehistory leads on through scholasticism and humanism to the academies of mathematicians of the seventeenth and eighteenth centuries and on to the Romantic philosophies and the runes of Novalis's hallucinatory visions. This same eternal idea, which for us has been embodied in the Glass Bead Game, has underlain every movement of Mind toward the ideal goal of a *universitas litterarum*, every Platonic academy, every league of an intellectual elite, every *rapprochement* between the exact and the more liberal disciplines, every effort toward reconciliation between science and art or science and religion. Men like Abelard, Leibniz, and Hegel unquestionably were familiar with the dream of capturing the universe of the intellect in concentric systems, and pairing the living beauty of thought and art with the magical expressiveness of the exact sciences. In that age in which music and mathematics almost simultaneously attained classical heights, approaches and cross-fertilizations between the two disciplines occurred frequently. And two centuries earlier we find in Nicholas of Cues sentences of the same tenor: ["The Mind ... conforms itself to absolute possibility in order to measure all things with respect to their possibility. Likewise, it conforms itself to absolute necessity in order to measure all things in their oneness and simplicity, as does God. Likewise, it conforms itself to necessary connection in order to measure all things in their own being, and it conforms itself to determinate possibility in order to measure all things as they [actually] exist. Mind also measures symbolically and in a comparative way—for example, when it uses number and geometrical figures and makes itself to be a likeness of these."][2]

Incidentally, this is not the only one of Nicholas's ideas that almost seems to suggest our Glass Bead Game, or corresponds to and springs from a similar branch of the imagination as the play of thoughts which occurs in the Game. Many similar echoes can be found in his writings. His pleasure in

[2]The Winston's here give their own translation of Hesse's quote of the German translation of Nicholas's original Latin. For easier verification, I have substituted the English wording from Jasper Hopkins's *Nicholas of Cusa on Wisdom and Knowledge* (Minneapolis, MN: The Arthur J. Banning Press, 1996). The quote stems from chapter 9 of the section "The Layman on Mind," pp. 531-601: 569.

mathematics also, and his delight and skill in using constructions and axioms of Euclidean geometry as similes to clarify theological and philosophical concepts, likewise appear to be very close to the mentality of the Game.[3] At times even his peculiar Latin (abounding in words of his own coinage, whose meaning, however, was perfectly plain to any Latin scholar) calls to mind the improvisatory agility of the Game's language.

[... W]e suspect, although we cannot prove this by citations, that the idea of the Game also dominated the minds of those learned musicians of the sixteenth, seventeenth, and eighteenth centuries who based their musical composition on mathematical speculations. Here and there in the ancient literatures we encounter legends of wise and mysterious games that were conceived and played by scholars, monks, or the courtiers of cultured princes. These might take the form of chess games in which the pieces and squares had secret meanings in addition to their usual function. And of course everyone has heard of those fables and legends for the formative years of all civilizations which ascribe to music powers far greater than those of any mere art: the capacity to control men and nations. These accounts make of music a kind of secret regent, or a lawbook for men and their governments. From the most ancient days of China to the myths of the Greeks we find the concept of an ideal, heavenly life for men under the hegemony of music. The Glass Bead Game is intimately bound up with this cult of music ("in eternal transmutations the secret power of song greets us here below," says Novalis).[4]

(3: pp. 30-40, the chronicler in the center of his introduction)

We shall now give a brief summary of the beginnings of the Glass Bead Game. It appears to have arisen simultaneously in Germany and in England. In both countries, moreover, it was originally a kind of exercise employed by those small groups of musicologists and musicians who worked and studied in the new seminaries of musical theory. If we compare the original state of the Game with its subsequent developments and its present form,[5] it is much like comparing a musical score of the period before 1500, with its primitive notes and absence of bar lines, with an eighteenth-century score, let alone with one from the nineteenth with its confusing excess of symbols for dynamics, tempi, phrasing, and so on, which often made the printing of such scores a complex technical problem.

[3]See "The Layman on Mind," particularly chapter 6, pp. 551-555 in the Hopkins translation.

[4]These are lines 1-2 in sonnet II of "Zueignung" from Novalis's *Heinrich von Ofterdingen*.

[5]"Present" refers to the fictional chronicler's time of writing, a time around the year 2400.

The Game was at first nothing more than a witty method for developing memory and ingenuity among students and musicians. [... I]t was played both in England and Germany before it was "invented" here in the Musical Academy of Cologne, and was given the name it bears to this day, after so many generations, although it has long ceased to have anything to do with glass beads.

The inventor, Bastian Perrot of Calw, a rather eccentric but clever, sociable, and humane musicologist, used glass beads instead of letters, numerals, notes, or other graphic symbols. Perrot [...]found that the pupils at the Cologne Seminary had a rather elaborate game they used to play. One would call out, in the standardized abbreviations of their science, motifs or initial bars of classical compositions, whereupon the other had to respond with the continuation of the piece, or better still with a higher or lower voice, a contrasting theme, and so forth. It was an exercise in memory and improvisation quite similar to the sort of thing probably in vogue among ardent pupils of counterpoint in the days of Schütz, Pachelbel, and Bach—although it would then not have been done in theoretical formulas, but in practice on the cembalo, lute, or flute, or with the voice.

Bastian Perrot [...] constructed a frame, modeled on a child's abacus, a frame with several dozen wires on which could be strung glass beads of various sizes, shapes, and colors. The wires corresponded to the line of the musical staff, the beads to the time-values of the notes, and so on. In this way he could represent with beads musical quotations or invented themes, could alter, transpose, and develop them, change them and set them in counterpoint to one another. In technical terms this was a mere plaything, but the pupils liked it; it was imitated and became fashionable in England too. For a time the game of musical exercises was played in this charmingly primitive manner. And as is so often the case, an enduring and significant institution received its name from a passing and incidental circumstance. For what later evolved out of that students' sport and Perrot's bead-strung wires bears to this day the name by which it became popularly know, the Glass Bead Game.

A bare two or three decades later the Game seems to have lost some of its popularity among students of music, but instead was taken over by mathematicians. For a long while, indeed, a characteristic feature in the Game's history was that it was constantly preferred, used, and further elaborated by whatever branch of learning happened to be experiencing a period of high development or a renaissance. The mathematicians brought the Game to a high degree of flexibility and capacity for sublimation, so that it began to acquire something approaching a consciousness of itself and its possibilities.

[...] Having passed from the musical to the mathematical seminaries (a change which took place in France and England somewhat sooner than in Germany), the Game was so far developed that it was capable of expressing mathematical processes by special symbols and abbreviations. The players, mutually elaborating these processes, threw these abstract formulas at one another, displaying the sequences and possibilities of their science. This mathematical and astronomical game of formulas required great attentiveness, keenness, and concentration. Among mathematicians, even in those days, the reputation of being a good Glass Bead Game player meant a great deal; it was equivalent to being a very good mathematician.

At various times the Game was taken up and imitated by nearly all the scientific and scholarly disciplines, that is, adapted to the special fields. There is documented evidence for its application to the fields of classical philology and logic. [...] The visual arts soon followed suit, architecture having already led the way in establishing the links between visual art and mathematics. Thereafter more and more new relations, analogies, and correspondences were discovered among the abstract formulas obtained in this way. Each discipline which seized upon the Game created its own language of formulas, abbreviations, and possible combinations. Everywhere, the elite intellectual youth developed a passion for these Games, with their dialogues and progressions of formulas. The Game was not mere practice and mere recreation; it became a form of concentrated self-awareness for intellectuals. [...]

The astronomers, the classicists, the scholastics, the music students all played their Games according to their ingenious rules, but the Game had a special language and set of rules for every discipline and subdiscipline. It required half a century before the first step was taken toward spanning these gulfs. [...] It was the achievement of one individual which brought the Glass Bead Game almost in one leap to an awareness of its potentialities, and thus to the verge of its capacity for universal elaboration. And once again this advance was connected with music. A Swiss musicologist with a passion for mathematics gave a new twist to the Game, and thereby opened the way for its supreme development. This great man's name in civil life can no longer be ascertained; by this time the cult of personality in intellectual fields had already been dispensed with. He lives on in history as the Lusor (or also, Joculator) Basiliensis. [...] He invented for the Glass Bead Game the principles of a new language, a language of symbols and formulas, in which mathematics and music played an equal part, so that it became possible to combine astronomical and musical formulas, to reduce mathematics and music to a common denominator, as it were. [...]

After Joculator Basiliensis' grand accomplishment, the Game rapidly evolved into what it is today: the quintessence of intellectuality and art, the sublime cult, the *unio mystica* of all separate members of the *Universitas Litterarum*. [...]

For all that the Glass Bead Game had grown infinitely in technique and range since its beginnings, for all the intellectual demands it made upon its players, and for all that it had become a sublime art and science, in the days of Joculator Basiliensis it still was lacking in an essential element. Up to that time every game had been a serial arrangement, an ordering, grouping, and confronting of concentrated concepts from many fields of thought and aesthetics, a rapid recollection of eternal values and forms, a brief, virtuoso flight through the realms of the mind. Only after some time did there enter into the Game, from the intellectual stock of the educational system and especially from the habits and customs of the Journeyers to the East, the idea of contemplation.

This new element arose out of an observed evil. Mnemonists, people with freakish memories and no other virtues, were capable of playing dazzling games, dismaying and confusing the other participants by their rapid muster of countless ideas. In the course of time such displays of virtuosity fell more and more under a strict ban, and contemplation became a highly important component of the Game. [... T]here now arose the demand for a deeper and more spiritual approach. After each symbol conjured up by the director of a Game, each player was required to perform silent, formal meditation on the content, origin, and meaning of this symbol, to call to mind intensively and organically its full purport.

[...] Under the shifting hegemony of now this, now that science or art, the Game of games had developed into a kind of universal language through which the players could express values and set these in relation to one another. Throughout its history the Game was closely allied with music, and usually proceeded according to musical or mathematical rules. One theme, two themes, or three themes were stated, elaborated, varied, and underwent a development quite similar to that of the theme in a Bach fugue or a concerto movement. A Game, for example, might start from a given astronomical configuration, or from the actual theme of a Bach fugue, or from a sentence out of Leibniz or the Upanishads, and from this theme, depending on the intentions and talents of the player, it could either further explore and elaborate the initial motif or else enrich its expressiveness by allusions to kindred concepts. Beginners learned how to establish parallels, by means of the Game's symbols, between a piece of classical music and the formula for same law of nature. Experts and Masters of the Game freely wove the initial theme into unlimited combinations. For a long time one school of players favored the technique of stating side by side, developing in counterpoint, and finally harmoniously combining two hostile themes or ideas, such as law and freedom, individual and community. In such a Game the goal was to develop both themes or theses with complete equality and impartiality, to evolve out of thesis and antithesis the purest possible synthesis.

(4: pp. 81-82, the Music Master to Joseph Knecht)

Our mission is to recognize contraries for what they are: first of all as contraries, but then as opposite poles of a unity. Such is the nature of the Glass Bead Game. [...] Each of us is merely one human being, merely an experiment, a way station. But each of us should be on the way toward perfection. Remember this: one can be a strict logician or grammarian and at the same time full of imagination and music. One can be a musician or Glass Bead Game player and at the same time wholly devoted to rule and order.

(5: pp. 118-120, Joseph Knecht in a letter to a friend)

Let me remind you of the time the two of us, assigned to the same group, were so eagerly working on our first sketches for Glass Bead Games. Do you recall a certain day and a certain game? Our group leader had given us various suggestions and proposed all sorts of themes for us to choose from. We had just arrived at the delicate transition from astronomy, mathematics, and physics to the sciences of language and history, and the leader was a virtuoso in the art of setting traps for eager beginners like us and luring us on to the thin ice of impermissible abstractions and analogies. He would slip into our hands tempting baubles taken from etymology and comparative linguistics, and enjoyed seeing us grab them and come to grief. We counted Greek quantities until we were worn out, only to feel the rug pulled out from under us when he suddenly confronted us with the possibility, in fact the necessity, of accentual instead of quantitative scansion, and so on. [...] I suddenly realized that in the language, or at any rate in the spirit of the Glass Bead Game, [...]every symbol and combination of symbols led not hither and yon, not to single examples, experiments, and proofs, but into the center, the mystery and innermost heart of the world, into primal knowledge. Every transition from major to minor in a sonata, every transformation of a myth or a religious cult, every classical or artistic formulation was, I realized in that flashing moment, if seen with a truly meditative mind, nothing but a direct route into the interior of the cosmic mystery, where in the alternation between inhaling and exhaling, between heaven and earth, between Yin and Yang, holiness is forever being created.

[...] You will recall, at least in general outline, the Glass Bead Game exercise we constructed at that time, as pupils in the Third Course, and with the leader's assistance [...]. That game began with a rhythmic analysis of a fugal theme and in the center of it was a sentence attributed to Confucius. Now I am studying that entire game from beginning to end. That is, I am working through each of its phrases, translating it from the language of the Game back into its original language, into mathematics, ornament, Chinese, Greek, and so on.

(6: pp. 245, about Magister Knecht's inaugural *ludus solemnis*)

Ever since his stay in Mariafels, Knecht had been nurturing an idea for a Glass Bead Game which he now decided to use for his first ceremonial Game as Magister. The pretty idea was to base the structure and dimensions of the Game on the ancient ritual Confucian pattern for the building of a Chinese house: orientation by the points of the compass, the gates, the spirit wall, the relationships and functions of buildings and courtyards, their co-ordination with the constellations, the calendar and family life, and the symbolism and stylistic principles of the garden. Long ago, in studying a commentary on the *I Ching*, he had thought the mythic order and significance of these rules made an unusually appealing and charming symbol of the cosmos and of man's place in the universe. The age-old mythic spirit of the people in this tradition of domestic architecture had also seemed to him wonderfully and intimately fused with the mandarin and magisterial spirit of speculative scholarliness.

II Ancient Chinese Thinkers on Music

The Annals of Lü Buwei. The Almanacs, Book 5, Chapter 2: Great Music[6]

5/2.1 *The origins of music*
Lie in the distant past:
Born of measurement.
Founded by Grand One,
Grand One brought forth the Dyadic Couple;
The Dyadic Couple brought forth Yin and Yang.
Yin and Yang metamorphize and transform,
The one rising, the other falling,
Joined together in a perfect pattern.
Spinning and pulsing,
If dispersed, they rejoin,
And joined, disperse again.
This is called the "Invariable Principle of Nature."
Heaven and Earth turn like the wheel of a carriage.
Reaching the end, it begins again;
Reaching its limit, it reverts again,
Everything fitting the overall scheme.
Sun, moon, planets, and constellations:
Some move fast, others slow,
In the completion of their movements.
The four seasons alternately arise,
Some hot, others cold;
In some, the days are short; in others, long;
Yin and Yang metamorphize and transform,
Sometimes rising, sometimes falling,
Some times are soft, other times hard.
The myriad things that emerged
Were created by Grand One
And transformed by Yin and Yang.

[6]Italicized lines correspond to those Hesse excerpted for *The Glass Bead Game*, pp. 29-30. Note that the excerpt represents the Winstons' English translation of Hesse's quotation from Richard Wilhelm's 1920s German translation of the Chinese original. Necessarily, the wording differs somewhat from J. Knoblock and J. Riegel's 2000 American rendering.

When the youngest sprouts were first stipulated,
They were given shape through coagulation.
Shapes and forms had their hollow places,
So none was without its sound.
Sound was produced by harmonious union,
Harmonious union by agreement;
When the First Kings fixed their music,
They started from these principles.

5/2.2 *When the world was in Great Tranquility,*
The myriad things were peaceful and secure.
When all were in agreement with the proper standard,
Music could be perfect.
The technique for perfecting music
Lies in limiting appetites and desires.
When appetites and desires are not licentious,
Music can be developed.
The method for developing music
Must be emergence out of equilibrium
Equilibrium emerges from impartiality,
Impartiality from the Dao.

Thus only a man who has attainted the Dao can discourse on music!

It is not that doomed states and disgraced peoples lack music, but rather that their music does not convey joy. It is not that "a drowning man does not sing," or "a crazy man does not dance." The music of a disordered age is like these. When ruler and minister fail to keep their proper places, father and son fail in their proper duties, and husband and wife fail to maintain their proper relationship, the people groan and sigh; but can this be considered to be music?

5/2.3

As a general principle, music is the harmony between Heaven and Earth, and the perfect blend of Yin and Yang. What is present in a person from birth is the result of nature and requires no effort for a person to mast. What nature causes a man to desire, he cannot but seek after. What it causes a man to hate, he cannot but seek to avoid. Both his desires and aversions are received from nature, and the person plays no part in creating them, nor can he modify or alter them. Yet some scholars of the present age condemn music. What is the reason for this?

5/2.4

Great music brings delight, enjoyment, and pleasure to ruler and subject, father and son, and old and young alike. Delight and enjoyment are born of equilibrium, and equilibrium is born of the Dao. It is the nature of the Dao that when we look for it, it is invisible, and when we listen for it, it is inaudible, for it cannot be given material form. Whoever is aware of the visible in the invisible, the audible in the inaudible, and the form of the formless almost knows it. The Dao is the supreme instance of the seminal essence, for it cannot be given shape or name. "Forced to give it a name, I would call it 'Grand One."

[...]

The Annals of Lü Buwei. The Almanacs, Book 5, Chapter 3: Extravagant Music

5/3.1

All men use their lives to live, yet do not know what makes them live. All men use their faculty of understanding to know, yet they do not understand what makes them know. To understand how one comes to know is called "knowing the Dao." Not to understand how one comes to know is called "discarding the treasure." Anyone who discards the treasure is certain to meet with disaster. Many rulers of the present age regard pearls, jade, lances, and swords as treasures. The more the possess of these, the more their people resent them, the more their state is endangered, and the more uneasy is the person of the ruler. As a consequence, he will miss the true nature of "treasure."

The music of a disordered age corresponds to this. If its tones are made by instruments of wood and skin, their sound is like the sound of rumbling thunder. If its tones are made by instruments of metal and stone, their sound is like the sound of claps of thunder. If its tones are made by instruments of string and bamboo or by singers and dancers, their sound is like the loud shouting of a crowd. If these sounds are employed to shock the mental energies, startle the ears and eyes, and agitate the inborn nature, it is permissible. If they are used to create music, there will be no pleasure from it. Thus, the more extravagant is the music, the gloomier are the people, the more disordered the state, and the more debased its ruler. As a consequence, he will also miss the true nature of music.

5/3.2

As a general rule, what the ancient sage-kings esteemed in music was the pleasure it created. But Jie of the Xia and Zhou Xin of the Yin created extravagant music. In the sounds of gigantic drums, bells, lithophones, flutes and pipes, they regarded the large to be beautiful and the many to be wonderful. They first created strange and exotic things that ears have never before heard and eyes never before seen. They strove after ever greater effect, employing neither rule nor measure. When Song was in decline, the Thousand Bell set was cast; when Qui was in decline, the Great Regulator bell was cast; *when Chu was in decline, the "Shamanka Tones" were created. Extravagance produced further extravagance. When those who themselves possess the Dao examine such music, they see that it is missing the true nature of music.* "Missing the true nature of music" means that their music did not produce pleasure, and this is certain to produce injury. The relationship between their lives and their music is like that of ice to the blazing sun: instead of using it properly, they use it to attack themselves. *This is the result of not knowing the true nature of music and thus striving for extravagance.*

5/3.3

Music possessing a true nature is analogous to the corporeal body's possessing an inborn and essential nature. Where there is such an inborn and essential nature, it is imperative that it be nurtured. Cold and heat, work and ease, hunger and satiation—these six are not of themselves in balance. As a general principle nurturing involves overseeing which of the six is not in balance and then put it into balance. If over the long term one can maintain balance, then life will be prolonged. It is characteristic of living that the body of itself is absolutely quiescent; only after there has been some stimulus is there awareness, awareness that has been occasioned by the stimulus. If this continues and if there is no reversal, then one comes to be governed by one's appetites and desires. If one is governed by one's appetites and desires, one is certain to lose one's natural endowment. Moreover, if one's appetites and desires know no limit, then it is inevitable that greed, pettiness, sedition and rebellion will come to characterize one's thought and than wantonness, debauchery, profligacy, and wickedness will come to characterize one's undertakings. Thus,

> the strong will plunder the weak, the many tyrannize the few,
> the bold violate the timid, and the vigorous bully the frail.

All of these are the product of this.

The Annals of Lü Buwei. The Almanacs, Book 5, Chapter 4: Balanced Tones

[...]
5/4.3

In musical tones as well there are balances that should be kept. If they are too grand and large, the inner mind is unsettled. [...] If one is too small, the mind will feel cheated. [...] If musical tones are too clear, the inner mind will feel threatened. [...] If musical tones are too muddy, the inner mind will be depressed. [...] Thus, musical tones that are too grandiose, too small, too clear, or too muddy are all not kept in balance.

5/4.4

What does "kept in balance" mean? It means that the sounds are kept at the Mean. What is meant by "kept at the Mean?"

> The amplitude of its pitch should not exceed that produced by the string of the seven-foot tuner. The weight of the bell should not exceed the 12-catty stone.

This defines how amplitude of pitch and weight of the bell are kept in check. The note *gong* of "Yellow Bell" is the standard for tones and the basis for determining clarity and muddiness. Whatever is consistent with this standard is "kept in balance." If one listens to what is kept in balance while one's self has been kept in balance, the result will be harmony. When musical instruments commit no excess, this is the correct standard of balance and harmony.

Thus, the tones of an orderly age are peaceful and joyous because its policies are stable. The tones of a chaotic age are resentful and angry because its policies are perverse. The tones of a doomed state are sad and mournful because its policies are dangerous. It is a general principle that music is influenced by government and affected by customs. When customs are fixed, music adjusts itself to them, Thus in an age that possesses the Dao one has only to observe its music to know its customs, to observe its customs to know its government, and observe its government to know its ruler. The First Kings were, therefore, certain to rely on music as a means of professing their teachings.

III Various Instantiations of a Glass Bead Game

(a) The Thematic Material of a Discursive Game in Five Movements

Movement I: the chronicler's introduction
theme 1: the intellectual depravity of the Age of the Feuilleton
theme 2: the World of the Mind and its symbol, the glass bead game
motif 1: music
motif 2: Chinese wisdom
motif 3: the spirituality introduced by the Journeyers to the East
hidden thread: the quest for universal harmony

Movement II: two monologues and a dialogue (Knecht and Designori)
theme 1: the distance between the World of Mind and the world-at-large
theme 2: the desire for harmonization of the two worlds
motif 1: the need for mutual respect
motif 2: bitterness
motif 3: serenity

Movement III: Knecht's circular
theme 1: the World of the Mind, the dangers it ignores, the decline it faces
theme 2: the Age of the Feuilleton; learning from a past gone awry
motif 1: the ideal of service
motif 2: the need for an awareness of history
motif 3: the glass bead game as example of a non-utilitarian luxury

Movement IV: the Castalian Directorate's reply
theme 1: "sham harmony"
theme 2: distance as lamentable but inevitable and desirable
motif 1: awareness of history
motif 2: the ideal of service
motif 3: the relationship between individual and hierarchy

Movement V: young Josef Knecht's clandestine poems
theme 1: impermanence
theme 2: preservation through symbolic notation and its "performance"
motif 1: human potential
motif 2: service
motif 3: serenity

(b) Three Further Threads with Related Thematic Material

An imaginary glass bead game based on "the ancient ritual Confucian pattern for the building of a Chinese house"

theme 1: submission, readiness to serve as a member in a hierarchical order
theme 2: art as a treasure trove of cultural heritage
motif 1: body
motif 2: mind
motif 3: spirit

Insights gained under the perspective of the I Ching

theme 1: aloofness is a blemish
theme 2: smallness can be powerful and beneficial
motif 1: great men should seek to become sacred vessels
motif 2: learning is beneficial (in an attitude of humility and respect)
motif 3: disharmony and all rifts should be avoided

Joseph Knecht's five lives

theme 1: rift versus synthesis
theme 2: ancient Chinese wisdom: the power of smallness, learning with respect, service, hierarchy vs. individuality
motif 1: contemplation / meditation, with the goal of serenity
motif 2: ritual and art
motif 3: "calling," "awakening"; the successor who will hear the call

(c) The Life of the *Magister Ludi* Joseph Knecht

theme 1: longing for harmony, readiness to sacrifice career and life for it
theme 2: service, humility; learning from examples rather than doctrines
motif 1: meditation
motif 2: music as great love, glass bead game as profession since it combines art with ritual
motif 3: calls, doubts, awakening experiences; master–successor

Bibliography

Primary Sources

For Johannes Kepler's thoughts and theories, I have worked with the German and English translations of the three texts most relevant for my study, *Harmonices mundi*, *Mysterium cosmographicum*, and *Somnium seu opus posthumum de astronomia Lunari*:

- *Weltharmonik*, translated and commented by Max Caspar (Munich: R. Oldenbourg, 1936),
- *Das Weltgeheimnis*, translated and commented by Max Caspar (Munich: R. Oldenbourg, 1997),
- *Kepler's Traum vom Mond*, translated and commented by Ludwig Günther (Leipzig: B.G. Teubner, 1898),
- *The Harmony of the World*, translated and commented by E.J. Aiton and A.M. Duncan with J.V. Field (Philadelphia, PA: The American Philosophical Society, 1997),
- *The Secret of the Universe*, translated and commented by E.J. Aiton and A.M. Duncan with I. Bernard Cohen (Norwalk, CT: Abaris Books, 1999), and
- *Somnium: The Dream, or Posthumous Work on Lunar Astronomy*, translated and commented by Edward Rosen (Mineola, NY: Dover Publications, 2003).

For Hesse's work, I have relied on the recent editions: *Das Glasperlenspiel: Versuch einer Lebensbeschreibung des Magister Ludi Josef Knecht* (Frankfurt: Suhrkamp, [2]2002) and *The Glass Bead Game (Magister Ludi)*, translated by Richard and Clara Winston with a foreword by Theodore Ziolkowski (New York: Picador, 2002 [1969]).

My analysis of Hindemith's opera is based on the facsimile full score of *Die Harmonie der Welt: Oper in fünf Aufzügen*, which was kindly loaned me by B. Schott's Söhne, Mainz, as well as the libretto, separately published by B. Schott's Söhne in 1957. For the text in English, I quote (with a few emendations) from Anne Pehrisch's translation, published in the booklet accompanying the Wergo CD recording.

On Pythagoras and the Pythagoreans

Burkert, Walter, *Lore and Science in Ancient Pythagoreanism*, Cambridge, MA: Harvard University Press, 1972.

Delatte, Armand, *Études sur la littérature pythagoricienne*, Geneva: Slatkine, 1999.

—, *La vie de Pythagore de Diogène Laërce*, Hildesheim and New York: G. Olms, 1988.

Guthrie, William Keith Chambers, *A History of Greek Philosophy*, Cambridge, UK: Cambridge University Press, 1962- (especially vol. I).

Guthrie, Kenneth Sylvan, *The Pythagorean Sourcebook and Library : An Anthology of Ancient Writings which Relate to Pythagoras and Pythagorean Philosophy*, Grand Rapids: Phanes Press, 1988.

Hummel, Charles, *Pythagoras und die Meister von Chartres*, Bern: Kreis der Freunde um Hans Kayser, 1998.

Jahoda, Gerhard, *Identische Strukturen pythagoreischer Zahlenschemata*, Vienna: Lafite, 1971.

Levin, Flora Rose, *The Harmonics of Nicomachus and the Pythagorean Tradition*, University Park, PA: American Philological Association, 1975.

Münxelhaus, Barbara, *Pythagoras musicus: Zur Rezeption der pythagoreischen Musiktheorie als quadrivialer Wissenschaft im lateinischen Mittelalter*, Bonn-Bad Godesberg: Verlag für systematische Musikwissenschaft, 1976.

Nicomachos of Gerasa: Introduction to Arithmetic, trans. Martin Luther d'Ooge, with studies in Greek arithmetic by F.E. Robbins and L.C. Karpinski, New York, NY: Macmillan, 1926.

Reis, Helmut, *Harmonie und Komplementarität: harmonikale Interpretation des pythagoreischen Lehrsatzes*, Bonn-Bad Godesberg: Verlag für Systematische Musikwissenschaft, 1983.

Stössel, Rudolf, *Kleine Einführung in die Pythagoreische Harmonik*, Hedingen: A. Scheidegger, 1977.

van der Waerden, Bartel Leendert, *Die Pythagoreer: Religiöse Bruderschaft und Schule der Weisheit*, Zurich/Munich: Artemis, 1979.

Vogel, Martin, *Die Enharmonik der Griechen*, Düsseldorf: Gesellschaft zur Förderung der systematischen Musikwissenschaft, 1963.

West, M.L., *Early Greek Philosophy and the Orient*, Oxford, UK: Clarendon Press, 1971.

On the harmonic order in the universe

Colman, Samuel, *Harmonic Proportion and Form in Nature, Art and Architecture*, New York, NY: Dover, 1912.

Goldschmidt, Victor, *Über Harmonie und Complication*, Berlin: J. Springer, 1901.

—, *Über Complication und Displication*, Heidelberg: C. Winter, 1921.

—, "Über Harmonie im Reich der Planetoiden," Leipzig: Akademische Verlagsgesellschaft, 1912.

—, Krystallographische Winkeltabellen, Berlin: Springer, 1897.

—, Materialien zur Musiklehre, Heidelberg : Winter, 1925, 2 volumes
.

Haase, Rudolf, *Aufsätze zur Geschichte der Harmonik: Platon, Leibniz, Bahr, Hauer, Hesse*, Bern: Kreis der Freunde von Hans Kayser, 1984.

—, *Das pythagoreische Lambdoma in neuer Sicht*, Vienna: Österreichische Zeitschrift für Musik (offprint), 1962.

—, *Geschichte des harmonikalen Pythagoreismus*, Vienna: Lafite, 1969.

—, *Grundlagen der harmonikalen Symbolik*, Munich: ORA Verlag, 1966.

—, *Harmonikale Synthese*, Vienna: Lafite, 1980.

—, *Leibniz und die Musik*, Hommerich: Eckhardt, 1963.

Iamblichus, *On the mysteries of the Egyptians, Chaldeans, and Assyrians and Life of Pythagoras, to which have been added ethical and political fragments of ancient Pythagorean writers*, trans. Thomas Taylor, Somerset, UK: Prometheus Trust, 1999.

Kayser, Hans, *Abhandlungen zur Ektypik harmonikaler Wertformen*, Zurich and Leipzig: M. Niehans, 1938.

—, *Akróasis: The Theory of World Harmonics*, Boston, MA: Plowshare Press, 1970.

—, *Der hörende Mensch*, Stuttgart: Engel, 1993.

—, *Die Form der Geige, aus dem Gesetz der Töne gedeutet*, Zurich: Occident-Verlag, 1947.

—, *Die Harmonie der Welt*, Zurich: Rhein-Verlag, 1959.

—, *Ein harmonikaler Teilungs-Kanon: Analyse einer geometrischen Figur im Bauhüttenbuch Villard de Honnecourts*, Zurich: Occident, 1946.

—, *Grundiss eines Systems der harmonikalen Wertformen*, Zurich: Occidentverlag, 1946.

—, *Harmonia plantarum*, Basel: Schwabe, 1943.

—, *Lehrbuch der Harmonik*, Zurich: Occident-Verlag, 1950.

—, *Orphikon: Eine harmonikale Symbolik*, Basel: Schwabe, 1973.

Kepler, Johannes, *Die Zusammenklänge der Welten*, ed. W. Harburger, Jena: E. Diederichs, 1918.

—, *Kosmische Harmonie*, ed. W. Harburger, Leipzig: Insel, 1925.

Kolk, Dieter, *Zahl und Qualität: Abhandlungen zur Harmonik Hans Kaysers*, Bern: Kreis der Freunde von Hans Kayser, 1995

McClain, Ernest, *The Myth of Invariance: The Origin of the Gods, Mathematics and Music from the* Rg Veda *to Plato*, New York, NY: Nicolas Hays, 1976.

Spitzer, Leopold, *Die harmonikale Symbolik des Albert von Thimus*, Vienna: Lafite, 1978

Studer, André, *Kriterien einer integralen Architektur; Werk und Transzendenz; Von der Idee zur Gestalt*, Bern: Kreis der Freunde von Hans Kayser, 1984.

—, *Architektur, Mensch, Mass*, Bern: Kreis der Freunde von Hans Kayser, 1976.

Strohmeier, John, *Divine Harmony*, Berkeley, CA: Hi Marketing, 2002.

Thimus, Albert von, *Die harmonikale Symbolik des Alterthums*, Hildesheim and New York: G. Olms, 1988.

Zipp, Friedrich, *Vom Urklang zur Weltharmonie: Werden und Wirken der Idee der Sphärenmusik*, Kassel: Merseburger, 1998.

On Kepler

Caspar, Max, *Johannes Kepler*, Stuttgart: Verlag für Geschichte der Naturwissenschaften und der Technik, 1995.

Cassirer, Ernst, *The Individual and the Cosmos in Renaissance Philosophy*, Mineola, NY: Dover Publications, 2000.

Gingerich, Owen, *The Eye of Heaven: Ptolemy, Copernicus, Kepler*, New York, NY: American Institute of Physics, c1993.

Jardine, Nicholas, *The Birth of History and Philosophy of Science: Kepler's* Defence of Tycho against Ursus, *with Essays on its Provenance and Significance*, Cambridge and New York: Cambridge University Press, 1988.

Lemcke, Mechthild, *Johannes Kepler*, Reinbek: Rowolt, 1995.

Lombardi, Anna Maria, *Kepler, le musicien du ciel*, Paris: Pour la science, 2003.

Martens, Rhonda, *Kepler's Philosophy and the New Astronomy*, Princeton, NJ: Princeton University Press, 2000.

Voelkel, James R., *Johannes Kepler and the New Astronomy*, New York, NY: Oxford University Press, 1994.

On Hindemith's Opera *The Harmony of the World*

Briner, Andres, "Eine Bekenntnisoper Paul Hindemiths: Zu seiner Oper *Die Harmonie der Welt*," in *Schweizerische Musikzeitung* 99 (1959), pp. 1-5 and 50-56.

— , "Die erste Textfassung von Paul Hindemiths Oper *Die Harmonie der Welt*," in Carl Dahlhaus, ed., *Festschrift für einen Verleger. Ludwig Strecker zum 90. Geburtstag*, Mainz: B. Schott's Söhne, 1973, pp. 203-241.

d'Angelo, James, "Tonality and symbolism in Hindemith's opera *Die Harmonie der Welt*," in *Hindemith-Jahrbuch* XIV (1985), pp. 99-128.

Gerstenberg, Walter, "Hindemiths Oper *Die Harmonie der Welt*: Historische und musikalische Perspektiven," in *Programmheft zur Uraufführung* (Munich 11 August 1957), pp. 34-36.

Haase, Rudolf, "Von Keplers 'Weltharmonik' zu Hindemiths 'Harmonie der Welt'," in *Österreichische Musikzeitschrift* 7-8 (1980), pp. 380-385.

—, *Paul Hindemiths harmonikale Quellen: Sein Briefwechsel mit Hans Kayser*, Vienna: Lafite, 1973.

—, "Johannes Kepler," in *Musik in Geschichte und Gegenwart*, Kassel: Bärenreiter/Metzler, ...

On Hesse and his *Glass Bead Game*

Arnold, Heinz Ludwig, *Hermann Hesse*, Munich: edition text und kritik 10-11, 1983.

Böhme, Wolfgang, *Suche nach Einheit: Hermann Hesse und die Religionen*, Frankfurt: Otto Lembeck, 1978.

Boulby, Mark, *Hermann Hesse: His Mind and Art*, Ithaca, NY: Cornell University Press, 1967.

Bran, Friedrich and Martin Pfeifer, *Hermann Hesse und die Religion*, Bad Liebenzell: Verlag B. Gegenbach, 1990.

Carlsson, Anni, "Hermann Hesses 'Glasperlenspiel' in seinen Wesensgesetzen," in Volker Michels, *Materialien, 'Das Glasperlenspiel'*, vol. II, pp. 39-59.

Chi, Ursula, *Die Weisheit Chinas und 'Das Glasperlenspiel'*, Frankfurt: Suhrkamp, 1976.

Dorner, Leo, "Glasperlenspiel und Musik: Eine musikphilosophische Deutung," in M. Pfeifer and F. Bran, eds., *Hermann Hesses Glasperlenspiel*, Bad Liebenzell: Verlag B. Gegenbach, 1987, pp. 91-98.

Field, G. Wallis, *Hermann Hesse – Kommentar zu sämtlichen Werken*, Stuttgart: H.D.Heinz, 1977.

Gellner, Christoph, *Weisheit, Kunst und Lebenskunst: fernöstliche Religion und Philosophie bei Hermann Hesse und Bertolt Brecht*, Mainz: Matthias-Grünewald-Verlag, 1994.

Grimes, James Milton, *The Nature of Hesse's Glass Bead Game and its Funtion in the Novel* Das Glasperlenspiel, Dissertation Vanderbilt, 1972.

Hsia, Adrian, *Hermann Hesse und China: Darstellung, Materialien und Interpretation*, Frankfurt: Suhrkamp, 1974.

— (ed.), *Hermann Hesse heute,* Bonn: Bouvier, 1980.

—, "Das esoterische Glasperlenspiel," in Volker Michels, ed., *Materialien zu Hermann Hesse 'Das Glasperlenspiel'*, vol. II, pp. 193-203.

—, "I Ging und das Glasperlenspiel Hesses,"in Arnold, ed., *Hermann Hesse*, pp. 53-66 [59].

Jacobs, Jürgen, *Wilhelm Meister und seine Brüder: Untersuchungen zum deutschen Bildungsroman,* Munich: W. Fink, 1972.

Karalaschwili, Reso, "Josef Knecht's Tod," in Volker Michels, ed., *Materialien zu Hermann Hesse 'Das Glasperlenspiel'*, vol. II, pp. 220-273.

Kirchhoff, Gerhard, "Josef Knechts Leben aus dem Geist der Musik: Eine Annäherung an das 'Glasperlenspiel'," in M. Pfeifer and F. Bran, eds., *Hermann Hesses Glasperlenspiel*, Bad Liebenzell: Verlag B. Gegenbach, 1987, pp. 68-90.

Lee, Inn-Ung, "I Ging, das Buch der Wandlungen im 'Glasperlenspiel'," in Eva Zimmermann, ed., *"Der Dichter sucht Verständnis und Erkanntwerden": Neue Arbeiten zu Hermann Hesse und seinem Roman* Das Glasperlenspiel, Bern: Peter Lang, 2002, pp. 53-71.

Mayer, Gerhart, "Christentum und asiatische Mystik im Werk Hermann Hesses," in Böhme, ed., *Suche nach Einheit*, pp. 25-54.

Michel, Willy und Edith Michel, "*Das Glasperlenspiel*," in Reclam-Interpretationen: *Hermann Hesse. Romane*, Stuttgart: Reclam, 1994, pp. 132-168.

Michels, Volker, "Zur Entstehung des 'Glasperlenspiels'," in *Hermann Hesse*: text und kritik, pp. 43-52 [49].

Michels, Volker, ed., *Materialien zu Hermann Hesse 'Das Glasperlenspiel'*, Frankfurt: Suhrkamp, 1973-74, 2 volumes.

Mileck, Joseph, *Hermann Hesse*'s Glasperlenspiel, Berkeley: University of California Press, 1952.

—, *Hermann Hesse: Life and Mind*, Berkeley: University of California Press, 1978.

Muschg, Adolf, "Hesses *Glasperlenspiel*," in Eva Zimmermann, ed., *Der Dichter sucht Verständnis ...*, pp. 125-138

Norton, Roger C., *Hermann Hesse's Futuristic Idealism:* The Glass Bead Game *and its Predecessors*, Frankfurt: Lang, 1973.

Pfeifer, Martin, *Hesse – Kommentar zu sämtlichen Werken*, Munich: Winkler, 1980.

Philippi, Klaus-Peter, "Hermann Hesse, *Das Glasperlenspiel*: 'Zerfall der Werte' und Flucht in die Legende," in Hermann Hesse, 1877-1962-2002, Tübingen: Attempto-Verlag, 2003.

Röttger, Jörg, *Die Gestalt des Weisen bei Hermann Hesse*, Bonn: Bouvier, 1980.

Thele, Andreas, *Hermann Hesse und Elias Canetti im Lichte ostasiatischer Geistigkeit*, Dissertation Düsseldorf 1993.

Tusken, Lewis W., *Hermann Hesse: The Man, His Myth, His Metaphor*, Columbia, SC: University of South Carolina Press, 1998.

Zhuang Yin Chen, *Asiatisches Gedankengut im Werke Hermann Hesses*, Bern: Peter Lang, 1997.

Zimmermann, Eva, ed., *"Der Dichter sucht Verständnis und Erkanntwerden": Neue Arbeiten zu Hermann Hesse und seinem Roman* Das Glasperlenspiel, Bern: Peter Lang, 2002.

Ziolkowski, Theodore, *The Novels of Hermann Hesse: A Study in Theme and Structure*, Princeton, NJ: Princeton University Press, 1965.

On Hesse's *Glass Bead Game* in light of the harmonic tradition

Gruntz-Stoll, Johannes, *Harmonik, Sprache des Universums: Überlieferung und Überwindung pythagoräischer Harmonik,* Bern: Kreis der Freunde um Hans Kayser, 2000.

Haase, Rudolf, "Hermann Hesse und die harmonikale Tradition: Eine Studie zu den Quellen des 'Glasperlenspiels'," *Neue Zürcher Zeitung* (12/13 October 1974), reprinted in *Aufsätze zur Geschichte der Harmonik: Platon, Leibniz, Bahr, Hauer, Hesse*, Bern: Kreis der Freunde von Hans Kayser, 1984, pp. 65-71.

Neubäcker, Peter, ed. *Harmonik und Glasperlenspiel: Beiträge 1993*, Munich: Neubäcker, 1994.

On games and attempts to construct a "glass bead game"

Aitkens, Dunbar, *Glass Plate Game: A Cooperative Thinking Game for Three to Ten Players*, Corvalis (microform), 1978.

Burk, Lucia, *Prognos: Glasperlen-Spiel, Glasperlen-Orakel*. Burgdorf: (n.d.)

Elffers, Joost, *Tangram: Das alte chinesische Formenspiel*, Cologne, Dumont, 1978.

Bayerische Akademie der Schönen Künste, *Der Mensch und das Spiel in der verplanten Welt*, Munich: Deutscher Taschenbuch Verlag, 1976..

Goldschmidt, Hannelore, *Globulus Cusani: Das Kugelspiel des Nikolaus von Kues*, Trier: Paulinus-Verlag, 1989.

Huizinga, Johan, *Homo ludens*, Boston, MA: Beacon Press, 1960.

Pennington, George, *Little manual for players of the glass bead game: the way of visual contemplation*, Tisbury, UK: Element Books, 1983.

On Chinese House Building

Thilo, Thomas, *Klassische chinesische Baukunst: Strukturprinzipien und soziale Funktion*, Zürich: Buchclub Ex Libris, 1978.

Klaas Ruitenbeek, *Carpentry and Building in Late Imperial China: A Study of the Fifteenth-Century Carpenter's Manual* Lu Ban jing (Leiden: Brill, 1993).

List of Musical Examples, Figures, and Illustrations

Musical Examples

Figures

Plates

Index

About the Author

Siglind Bruhn is a musicologist, concert pianist, and interdisciplinary scholar whose research focuses on compositions of the 20th century. Prior to coming to the United States, she taught for ten years in Germany and at the University of Hong Kong. Since 1993 she has been a full-time researcher at the University of Michigan's Institute for the Humanities (one of six "Life Research Associates"); in the fall of 2004, she was appointed *chercheur permanent* at the Institut d'Esthétique des Arts Contemporains at Université de Paris 1–La Sorbonne. She has been en elected member of the European Academy of Arts and Sciences since 2001.

Siglind Bruhn is the author of fourteen book-length monographs and numerous scholarly essays as well as the contributing editor of four essay collections. In the course of the past ten years, she published:

Saints in the Limelight: Representations of the Religious Quest on the Post-1945 Operatic Stage (Pendragon, 2003), Musical Ekphrasis: Composers Responding to Poetry and Painting (Pendragon, 2000), Musical Ekphrasis in Rilke's Marienleben (Rodopi, 2000), *The Temptation of Paul Hindemith:* Mathis der Maler *as a Spiritual Testimony* (Pendragon, 1998), *Images and Ideas in Modern French Piano Music* (Pendragon, 1997);

Christus als Opernheld im späten 20. Jahrhundert (Gorz, 2005), *Das tönende Museum: Musik des 20. Jahrhunderts interpretiert Werke bildender Kunst* (Gorz, 2004), *Musikalische Symbolik in Olivier Messiaens Weihnachtsvignetten* (Peter Lang, 1997).

Edited volumes: *Voicing the Ineffable: Musical Representations of Religious Experience* (Pendragon, 2002), *Signs in Musical Hermeneutics* (special issue of *The American Journal of Semiotics*, 1998), *Encrypted Messages in Alban Berg's Music* (Garland, 1998), *Messiaen's Language of Mystical Love* (Garland, 1998).